Advanced Biologically Inspired Vision and Its Application

Advanced Biologically Inspired Vision and Its Application

Guest Editor

Jie Cao

MDPI

Basel • Beijing • Wuhan • Barcelona • Belgrade • Novi Sad • Cluj • Manchester

Guest Editor
Jie Cao
School of Optics and
Photonics
Beijing Institute of
Technology
Beijing
China

Editorial Office
MDPI AG
Grossspeteranlage 5
4052 Basel, Switzerland

This is a reprint of the Special Issue, published open access by the journal *Biomimetics* (ISSN 2313-7673), freely accessible at: https://www.mdpi.com/journal/biomimetics/special_issues/FYI75A0551.

For citation purposes, cite each article independently as indicated on the article page online and as indicated below:

Lastname, A.A.; Lastname, B.B. Article Title. *Journal Name* **Year**, *Volume Number*, Page Range.

ISBN 978-3-7258-5303-8 (Hbk)
ISBN 978-3-7258-5304-5 (PDF)
https://doi.org/10.3390/books978-3-7258-5304-5

Cover image courtesy of Jie Cao

Contents

About the Editor . . . . **vii**

Xuan Xiao, Mingming Su, Bailiang Guo, Jingxue Wu, Jianming Wang and Jiayu Liang
Design and Experimental Validation of Pipeline Defect Detection in Low-Illumination Environments Based on Bionic Visual Perception
Reprinted from: *Biomimetics* **2025**, *10*, 569, https://doi.org/10.3390/biomimetics10090569 . . . . **1**

Hyuk Jae Jang, Jaemin Jeon, Joo Ho Yun, Iqbal Shudha Tasnim, Soyeon Han, Heeyoung Lee, et al.
Dual-Coated Antireflective Film for Flexible and Robust Multi-Environmental Optoelectronic Applications
Reprinted from: *Biomimetics* **2024**, *9*, 644, https://doi.org/10.3390/biomimetics9100644 . . . . . **25**

Bin Han, Quanchao Zhao, Moudan Shi, Kexin Wang, Yunan Shen, Jie Cao, et al.
Eye-Inspired Single-Pixel Imaging with Lateral Inhibition and Variable Resolution for Special Unmanned Vehicle Applications in Tunnel Inspection
Reprinted from: *Biomimetics* **2024**, *9*, 768, https://doi.org/10.3390/biomimetics9120768 . . . . . **36**

Hedva Spitzer, Yosef Shai Kashi, Morris Mosseri and Jacob Erel
Visual System Inspired Algorithm for Enhanced Visibility in Coronary Angiograms (VIAEVCA)
Reprinted from: *Biomimetics* **2025**, *10*, 18, https://doi.org/10.3390/biomimetics10010018 . . . . . **49**

Bin Li, Yuki Todo and Zheng Tang
Artificial Visual System for Stereo-Orientation Recognition Based on Hubel-Wiesel Model
Reprinted from: *Biomimetics* **2025**, *10*, 38, https://doi.org/10.3390/biomimetics10010038 . . . . . **74**

Shangbo Liu, Baowang Lian, Jiajun Ma, Xiaokun Ding and Haiyan Li
Adaptive Neural Network Robust Control of FOG with Output Constraints
Reprinted from: *Biomimetics* **2025**, *10*, 372, https://doi.org/10.3390/biomimetics10060372 . . . . **96**

Xiaoge Wang, Yunlong Sheng, Qun Hao, Haiyuan Hou and Suzhen Nie
YOLO-HVS: Infrared Small Target Detection Inspired by the Human Visual System
Reprinted from: *Biomimetics* **2025**, *10*, 451, https://doi.org/10.3390/biomimetics10070451 . . . . **113**

Yiqun Zhang, Wu Wang, Xiaokun Ding, Liyu Sun, Zhenyang Qian, Huilin Jiang, et al.
The Effects of Turbulent Biological Tissue on Adjustable Anomalous Vortex Laser Beam
Reprinted from: *Biomimetics* **2025**, *10*, 461, https://doi.org/10.3390/biomimetics10070461 . . . . **128**

About the Editor

Jie Cao

Jie Cao is a tenured associate professor and doctoral supervisor at the School of Optics and Photonics, Beijing Institute of Technology, and the Yangtze River Delta Research Institute (Jiaxing) of Beijing Institute of Technology. He has long been committed to research in bionic vision and intelligent perception. Having broken through several key bottlenecks in optoelectronic imaging sensing technology, he has published more than 80 SCI/EI papers and obtained over 30 authorized invention patents in the field of bionics and imaging perception. His research achievements have been widely applied in national strategic key areas such as intelligent manufacturing, unmanned driving, and low-altitude economy, injecting new momentum of intelligent vision into industrial development.

biomimetics

MDPI

Article

Design and Experimental Validation of Pipeline Defect Detection in Low-Illumination Environments Based on Bionic Visual Perception

Xuan Xiao [1], Mingming Su [2], Bailiang Guo [3], Jingxue Wu [1], Jianming Wang [4] and Jiayu Liang [4,*]

1 School of Computer Science and Technology, Tiangong University, Tianjin 300387, China; xxiao@tiangong.edu.cn (X.X.); wujingxue1014@163.com (J.W.)
2 School of Software, Tiangong University, Tianjin 300387, China; 2331111311@tiangong.edu.cn
3 School of Electronics and Information Engineering, Tiangong University, Tianjin 300387, China; guobailiang@outlook.com
4 Tianjin Key Laboratory of Autonomous Intelligence Technology and Systems, Tiangong University, Tianjin 300387, China; wangjianming@tiangong.edu.cn
* Correspondence: liangjiayu@tiangong.edu.cn

Abstract

Detecting internal defects in narrow and curved pipelines remains a significant challenge, due to the difficulty of achieving reliable defect perception under low-light conditions and generating collision-free motion trajectories. To address these challenges, this article proposes an event-aware ES-YOLO framework, and develops a pipeline defect inspection experimental environment that utilizes a hyper-redundant manipulator (HRM) to insert an event camera into the pipeline in a collision-free manner for defect inspection. First, to address the lack of datasets for event-based pipeline inspection, the ES-YOLO framework is proposed. This framework converts RGB data into an event dataset, N-neudet, which is subsequently used to train and evaluate the detection model. Concurrently, comparative experiments are conducted on steel and acrylic pipelines under three different illumination conditions. The experimental results demonstrate that, under low-light conditions, the event-based detection model significantly outperforms the RGB detection model in defect recognition rates for both types of pipelines. Second, a pipeline defect detection physical system is developed, integrating a visual perception module based on the ES-YOLO framework and a control module for the snake-like HRM. The system controls the HRM using a combination of Nonlinear Model Predictive Control (NMPC) and the Serpentine Crawling Algorithm (SCA), enabling the event camera to perform collision-free inspection within the pipeline. Finally, extensive pipeline insertion experiments are conducted to validate the feasibility and effectiveness of the proposed framework. The results demonstrate that the framework can effectively identify steel pipeline defects in a 2 Lux low-light environment, achieving a detection accuracy of 84%.

Keywords: bio-inspired event vision; ES-YOLO; deep learning with Event data; synthetic event data; hyper-redundant manipulator; NMPC

1. Introduction

Pipelines typically present a narrow, complex, and poorly illuminated environment [1], making manual inspection of pipeline defects a considerable challenge. This challenge arises primarily from two factors. First, inadequate lighting conditions hinder sensors from clearly capturing the internal pipeline environment. Second, the narrow and curved geometry of pipelines restricts the effectiveness of traditional manual inspection and remotely operated robotic systems, which are not only inefficient and vulnerable to environmental disturbances but also struggle to detect concealed defects. In recent years, with advancements in sensor technologies and intelligent robotics, pipeline inspection has increasingly evolved toward automation and intelligence.

Visual perception is widely applied in pipeline defect detection. Le et al. [2] developed an online multi-sensor inspection system for pipeline defect detection, employing RGB fusion technology to precisely identify defects such as cracks, corrosion, and blockages. Nevertheless, the limitations of RGB cameras in low-light and high-dynamic-range environments remain a significant challenge in complex pipeline inspections.

In recent years, event cameras, as bio-inspired visual sensors, have become increasingly important in industrial inspection due to their advantages in dynamic range, temporal resolution, low power consumption, and high sensitivity to low-light conditions [3]. This makes event cameras particularly suited for low-light and high-contrast environments. However, existing computer vision algorithms are not directly compatible with event streams, posing challenges for the direct application of event cameras in data processing. In addition, the data format of RGB and event streams differs significantly, and most existing standard datasets are based on RGB images, with a lack of publicly available datasets specifically designed for event data. This further restricts the application and development of event cameras in specific fields. Recreating pipeline defect datasets suitable for event cameras is not only time-consuming but also costly.

To address this issue, Henri Rebecq et al. [4] designed the event camera simulator ESIM. Building on this, Daniel Gehrig et al. [3] introduced frame upsampling technology into this framework, allowing models trained on synthetic event data to better generalize to real event data. Although these methods offer potential for applying event cameras to pipeline defect detection, whether the model can preserve the advantages of event cameras in low-light conditions remains a critical issue. Furthermore, beyond visual perception, ensuring that a pipeline inspection robot can navigate a collision-free trajectory for safe insertion into the pipeline constitutes an additional significant challenge. Currently, pipeline robots have been developed in various structures and configurations [5–7]. These technologies offer significant advantages in improving detection accuracy and reducing labor costs.

Among them, snake-like hyper-redundant manipulators (HRMs), due to their high flexibility, adaptability, and multiple degrees of freedom, can operate efficiently in complex pipeline environments. However, a key challenge in their path planning lies in ensuring that the end-effector reaches the target position while avoiding collisions between the manipulator and its surroundings. Although the high redundancy enhances the manipulator's reachability and spatial exploration capabilities, it also significantly increases the computational time and resource consumption of path planning [8,9]. The E-RRT* algorithm proposed by Ji et al. [10] employed ellipses instead of traditional straight lines to connect path nodes and optimized the sampling process, effectively addressing the path planning problem for HRMs in confined spaces.

Due to their slender and flexible bodies, snakes efficiently explore narrow spaces, making them a key inspiration for bionic design. Inspired by this, previous studies have proposed modular snake-like structures to construct HRMs [11]. Furthermore, snake locomotion exhibits a "head-following" characteristic, whereby body movements are determined by the trajectory of the head. This mechanism can be applied to the motion planning of snake-like robots: if the final configuration of the robot's end-effector reaching the target point is directly planned, this configuration can be regarded as the head trajectory, and the body joints follow this trajectory. Zhu et al. [12] proposed a pipeline inspection method that combines Nonlinear Model Predictive Control (NMPC) with a Serpentine-inspired Crawling Algorithm (SCA).

In summary, this article proposes and validates an innovative framework, ES-YOLO, designed to address the core challenges of employing event cameras for defect detection in narrow, low-light pipelines. To this end, we first present a practical method to systematically convert a public RGB dataset (NEU-DET) into event-based data, thereby creating a novel dataset, N-neudet. This dataset is then utilized to train an efficient defect detection model. Furthermore, the proposed perception framework is integrated with the motion control system of a snake-like hyper-redundant manipulator, as illustrated in Figure 1. The feasibility and practical effectiveness of the overall system are validated through physical experiments conducted in pipelines of different materials and under extremely low-light conditions. The main contributions of this article are as follows.

- To the best knowledge of the authors, this work presents the first system-level integration and validation of event-driven perception with an HRM for pipeline inspection, thereby constructing and experimentally verifying a complete HRM-based defect detection platform. The contribution goes beyond algorithmic testing under idealized conditions by providing a holistic validation of the entire perception and motion framework under real-world physical constraints. These results convincingly demonstrate both the feasibility and the significant potential of this technical approach for addressing complex industrial environments.
- An event-based defect dataset, termed N-neudet, is successfully constructed, with its data synthesized from traditional RGB images using the ES-YOLO framework. The practical utility of this dataset is rigorously validated through comprehensive comparative experiments. A defect detection system trained on N-neudet exhibits remarkable stability and accuracy, particularly under low-light conditions where conventional RGB cameras proved ineffective. These results confirm the robustness of the dataset and its suitability for event-based pipeline defect detection applications.

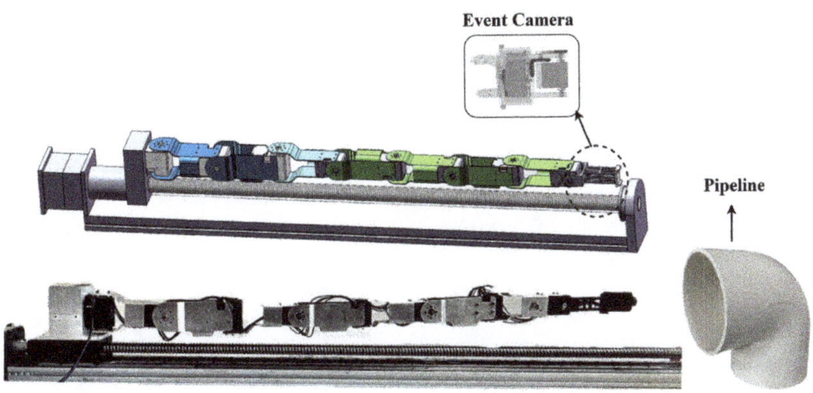

Figure 1. Schematic diagram of the model of an HRM equipped with an event camera.

2. Related Work

2.1. Trajectory Planning for HRM

Due to their high redundancy, HRMs can flexibly avoid obstacles and effectively access narrow, curved spaces.

The nonlinear control of HRMs poses significant challenges due to their high degrees of redundancy. Liu et al. [13] addressed this issue by proposing a novel control framework based on a nonlinear observer, which transformed the system into an interconnected input-to-state stable (ISS) structure, thereby achieving asymptotic stability. Building upon this, Liu et al. [14] employed the small-gain theorem to design distributed optimal controllers that effectively resolved the output consensus problem in multi-agent nonlinear systems. More recently, Jin [15] tackled the global asymptotic stability problem of feedback optimization in nonlinear systems by integrating an enhanced gradient flow optimizer with a nonlinear perturbation function. The proposed method was rigorously validated using tools such as singular perturbation theory and input-to-state stability analysis.These studies provide theoretical guidance and inspiration for the nonlinear control of HRMs.

Currently, prevalent path planning methods primarily include graph search-based algorithms and sampling-based algorithms. Among these, the A* algorithm [16–18] and Dijkstra's algorithm [9] were common representatives of graph search methods. Tang et al. [19] enhanced the A* algorithm by incorporating the artificial potential field method, introducing new node search strategies, and integrating local path optimization. This approach effectively reduced the number of search nodes, improved search efficiency, and simultaneously optimized the manipulator's obstacle avoidance posture. Conversely, ref. [20] proposed an extended Dijkstra's algorithm that integrates Delaunay triangulation and plane transformation techniques to optimize paths on complex surfaces, significantly enhancing path planning accuracy in both single-robot and multi-robot tasks.

Ref. [12] proposed a pipeline inspection and detection method that integrates Nonlinear Model Predictive Control (NMPC) with a Snake-inspired Heuristic Crawling Algorithm (SCA). The method is divided into three stages: insertion, inspection, and exit. During the insertion and withdrawal stages, the SCA—inspired by snake locomotion—is employed to significantly reduce path planning time. In the inspection stage, NMPC is applied to ensure efficient and collision-free configuration. Specifically, NMPC functions as a high-level planner, generating a global, collision-free path configuration. In contrast, SCA operates as a low-level controller, guiding the manipulator to accurately track the sequence of path points generated by NMPC, thereby executing the physical insertion motion.

2.2. Deep Learning with Event Data

In the field of event cameras, significant efforts have been made to create datasets and explore their applications, particularly in object detection.

For example, in [21], the authors used a Spiking Neural Network (SNN) to simulate human eye saccadic movements, converting static image datasets into event data, thereby creating semi-synthetic event-based versions of the MNIST and Caltech101 datasets. In [22], the DVS-Gesture dataset, which includes 11 types of gestures, was used to achieve real-time recognition of gesture data streamed by DVS. Additionally, Enrico Calabrese et al. [23] created a 3D human pose dataset, and [24] created the active pixel vision sensors (DAVIS) Driving Dataset (DDD17), the first driving recording dataset combining DVS and DAVIS. Subsequently, ref. [25] proposed DSEC, a large-scale new dataset that includes data from high-resolution event cameras, traditional cameras, LiDAR, and GPS, and provides disparity ground truth, aimed at promoting and evaluating the development of event-based stereo vision algorithms.

These datasets have laid the foundation for advancements in object detection using event cameras. Cannici et al. [26] proposed two event camera-based object detection models, YOLE and fcYOLE. Maqueda et al. [27] designed a CNN architecture adapted to event camera outputs and proposed an event-frame representation method, successfully predicting steering angles in autonomous driving using the DDD17 [24] dataset based on event cameras. Subsequently, Alonso et al. [28] proposed the first semantic segmentation baseline model for the DDD17 dataset. However, event-based deep learning algorithms still lack support from large-scale event datasets.

2.3. Synthetic Event Data

Currently, there are many event datasets [21–23] available for event-driven algorithms. Rebecq et al. [29,30] made progress by using simulated event data to train a recurrent neural network-based model, which succeeded in video reconstruction tasks, demonstrating the potential for expanding event-camera applications. Kaiser et al. [31] proposed a simple event camera simulator that generates events based on image difference thresholds. Refs. [32,33] introduced two event generation simulators. Daniel Gehrig et al. [3] enhanced the event simulator ESIM [4] with frame interpolation technology, which can convert frame data from traditional video sequences into asynchronous event streams. This method allows for the reuse of existing datasets to generate event data and accurately simulates the behavior of real event cameras, ensuring that synthetic data is suitable for training and testing event algorithms.

3. Overall Workflow of the Pipeline Defect Detection Method

To evaluate a performance of the ES-YOLO framework in a real-world pipeline environment, a dedicated experimental platform for defect detection is developed. This platform integrates two core components: first, a visual perception module equipped with the ES-YOLO framework to address the challenge of detection in low-light environments; second, a snake-like HRM motion control system based on NMPC and SCA algorithms to achieve collision-free navigation within complex curved pipelines.

The experimental procedure is primarily divided into three steps.

- Step 1: Model Construction. The ESIM model is employed to synthesize an event-based dataset, designated N-neudet, from the public NEU-DET RGB dataset. This synthesized dataset is then utilized to train a YOLOv8 model, culminating in the final ES-YOLO framework.
- Step 2: Data Collection. An event camera is mounted on the HRM. Guided by the NMPC and SCA algorithms, the HRM first determines a collision-free trajectory and is subsequently inserted into the pipeline to collect in situ data.
- Step 3: Defect Detection and Evaluation. Upon completion of data acquisition, the resulting dataset is fed into the ES-YOLO framework for defect detection, and the performance is quantitatively evaluated.

The comprehensive experimental workflow is illustrated in Figure 2.

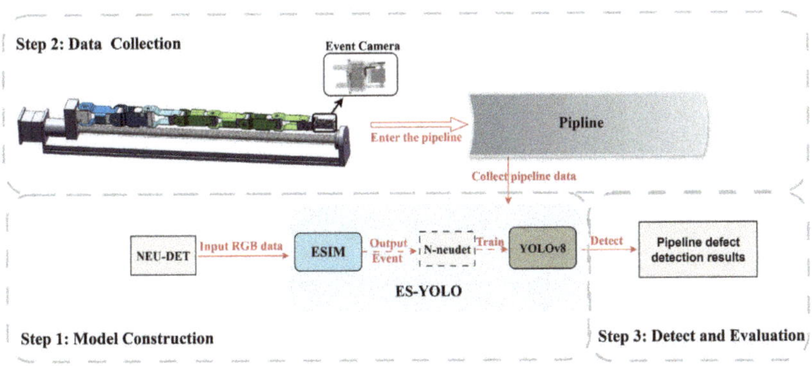

Figure 2. Overall workflow of the pipeline defect detection method based on the ES-YOLO.

4. Methodology of the ES-YOLO

4.1. Overview

The framework of ES-YOLO, which converts RGB datasets for event cameras, is illustrated in Figure 3.

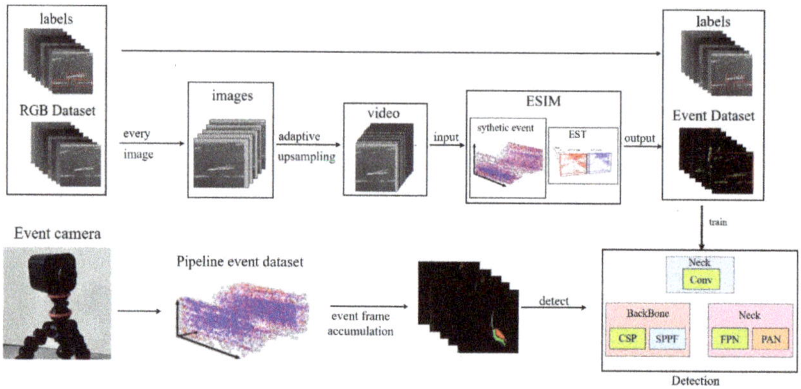

Figure 3. ES-YOLO framework architecture. First, large-scale image datasets (NEU-DET) are converted into synthetic event datasets (N-neudet) through the ESIM module; then, the generated event data is passed to the YOLOv8 module for training.

ES-YOLO consists of the ESIM module and the YOLOv8s module. This section introduces the ESIM module and the YOLOv8s module. First, the RGB dataset is upsampled to convert it into a high-frequency dataset. Second, the ESIM module is used to synthesize event dataset. Finally, the YOLOv8s model is trained using the N-neudet event dataset, and defect detection is performed on actual pipelines.

4.2. ESIM Module

ESIM module is used to synthesize event data. Due to the lack of large event datasets for defect detection, we need to convert RGB datasets into event datasets. Equation (1) [3,34,35] describes the event generation model of an ideal sensor. The pixels of an event camera are independent and continuously monitor the logarithmic brightness signal at their corresponding positions $L(\mathbf{u},t)$. When the change in logarithmic brightness $u = (x_i, y_i)^T$ at a pixel exceeds a certain threshold C over time t_i, an event $e_i = (x_i, y_i, t_i, p_i)$ is triggered.

$$\Delta L\left((x_i,y_i)^T,t_i\right) = L\left((x_i,y_i)^T,t_i\right) - L\left((x_i,y_i)^T,t_i-\Delta t_i\right) \geq p_i C \quad (1)$$

In Equation (1), $p_i \in \{-1, 1\}$ represents the polarity of the event signal, which is the indicator of the change in brightness. Δt represents the time since the last occurrence of pixel u.

The image is processed through an offset function to transform the original image into a low-frame-rate version. Although the ESIM event camera simulator can adaptively render virtual scenes at any temporal resolution, the actual event camera operates on a microsecond timescale. Therefore, we apply frame interpolation technology [36] to enable frame reconstruction at any temporal resolution. Simultaneously, an adaptive upsampling strategy [4] is employed to determine the number of intermediate frames to generate, thus converting the low-frame-rate images into a high-frame-rate sequence. Finally, the event synthesis module (ESIM) is used to convert the high-frame-rate image sequence into event data.

4.3. YOLOv8 Module

The event dataset generated by ESIM in Section 4.2 is then used as input for training the defect detection model. As mentioned in [37], YOLO is the first model to treat the object detection problem as a regression task. Released in 2023, YOLOv8 [38] builds upon the success of earlier versions such as YOLOv5 [37], introducing advanced architectural designs and training strategies, including anchor-free detection, to provide a unified framework with improved accuracy for various computer vision tasks [39,40]. By integrating an anchor-free design with attention mechanisms and dynamic convolution, the model not only simplifies its architecture but also significantly enhances the detection of small objects—an essential factor in many edge deployment scenarios. Given its excellent performance in terms of accuracy, efficiency, and usability, this study adopts YOLOv8 as the core detection module.

4.4. Event Spike Tensor

The synthetic events are combined with the original NEU-DET labels to train the YOLO module. Due to the sparsity of event signals and their non-uniform spatiotemporal distribution, pattern recognition algorithms typically aggregate event data into grid-based representations. To address this, we use the general framework Event Spike Tensor (EST) [41], which converts event streams into grid-based representations. Unlike previous event representation methods, EST retains the four dimensions of event signals and maps events of positive and negative polarities to two independent spatiotemporal grids, thereby enhancing the model's ability to represent event data.

5. Comparative Experiments Based on the NEU-DET Dataset

To validate the feasibility of the ES-YOLO framework for practical pipeline defect detection, we conducted a series of experiments. These experiments involved synthesizing event datasets, setting model training parameters, evaluating the model, and performing practical defect detection. Finally, we compared the detection results of the event camera with those of the RGB camera.

5.1. Dataset Generation

For the experiments, we selected the NEU-DET dataset [42], which contained six typical surface defects of hot-rolled strip steel: Roll Scale (RS), Patches (Pa), Cracks (Cr), Pitted Surface (PS), Inclusions (In), and Scratches (Sc). The dataset consisted of 1800 images, with 300 samples per defect type, and included detailed annotations specifying both the defect categories and their locations. After selecting the NEU-DET dataset, we processed it through the ESIM module to generate a synthetic event-based defect dataset, which we named N-neudet. This event dataset was subsequently used to train the YOLO-based defect detection model, allowing for the evaluation of the model's performance across different defect categories. As a result, a total of 1800 event defect images were generated, containing six different types of typical defects, with 300 samples for each defect type. The conversion process from RGB images to event frames is illustrated in Figure 4.

To improve the recognition accuracy, this article employed a sliding window combined with adaptive interpolation strategy. Specifically, a 512×512 Region of Interest (ROI) window was defined on a 640×640 static RGB image and was synchronously shifted in both horizontal and vertical directions with a stride of $s = 1$ pixel, resulting in $K = \left\lceil \frac{640-512}{s} \right\rceil + 1 = 129$ sub-regions (with the final incomplete region discarded). Each ROI sequence was synthesized into a 128-frame, 20 fps video, while its original label boundaries were mapped to the local coordinate system. This transformation was given by $\mathbf{x}' = \mathbf{x} - \Delta\mathbf{x}$, where $\mathbf{x}$ represented the bounding box coordinates in the original image,

$\Delta\mathbf{x}$ was the displacement vector of the ROI window's origin, and $\mathbf{x}'$ was the resulting coordinate vector in the local ROI frame. Subsequently, an adaptive frame interpolation technique [36] was applied to upsample the video from 20 fps to 200 fps, producing a high-frame-rate sequence. This sequence was then fed into the ESIM module to generate the corresponding event data. Compared to directly simulating events from the full 640 × 640 image, the proposed sliding window and adaptive interpolation approach significantly increased the target event density, thereby enhancing the recognition accuracy of the ES-YOLO framework.

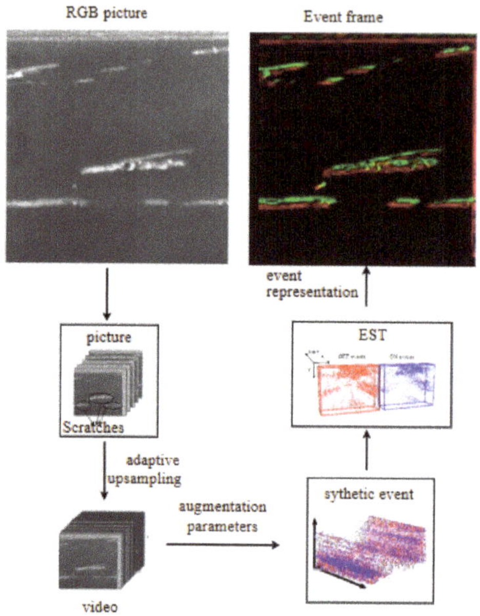

Figure 4. Illustration of RGB dataset conversed into event dataset.

To simulate event generation, we used the open-source simulator ESIM [3]. Before generating events, we needed to carefully select the contrast thresholds. We randomly sampled the contrast thresholds for positive (C_{tp}) and negative (C_{tn}) events from a uniform distribution, $U(C_{t\,\mathrm{min}}, C_{t\,\mathrm{max}})$. This random sampling method improved the domain adaptability between simulated and real data, thereby enhancing the generalization of simulated events to real events. In this experiment, we chose $C_{t\,\mathrm{min}} = 0.1$ and $C_{t\,\mathrm{max}} = 0.5$. After the ESIM simulation, the final event stream was generated with an event density of 256 events/ms and a temporal resolution of 1.953 ms.

Finally, to match the synthesized event stream with the annotations based on the YOLOv8 model, we converted the asynchronous and sparse event stream into a tensor representation. We selected the Event Spike Tensor (EST) [41] because it performed well in both high-level and low-level tasks. The EST method mapped positive and negative polarity events into separate spatiotemporal grids, which were then stacked along the channel dimension to form the final tensor representation of size $H \times W \times C$. In this representation, H and W were the spatial resolution of the sensor, and $C = 15$ was a hyperparameter which controls the number of temporal bins used to aggregate events.

5.2. Experimental Parameters

This experiment was implemented using the PyTorch framework. The ADAM optimizer was used for training, with a learning rate set to 0.01. A total of 100 epochs were trained, with the momentum parameter set to 0.937, and a batch size of 64. The training parameters used in the experiment are shown in Table 1.

Table 1. Training parameters.

Parameter	LR	Batch	Image	Monmentum	Epoch
Value	0.01	64	640*640	0.937	100

5.3. Evaluation Metrics

In this experiment, we used multiple metrics to evaluate the model's detection performance, including precision (P), recall (R), and mean average precision (mAP).

Precision is defined as the ratio of the number of True Positive samples correctly predicted by the model to the total number of samples predicted as positive by the model, as shown in Equation (2).

$$\text{Precision} = \frac{TP}{TP + FP} \tag{2}$$

Recall is defined as the ratio of the number of True Positive samples correctly predicted by the model to the total number of actual True Positive samples, as shown in the following Equation (3).

$$\text{Recall} = \frac{TP}{TP + FN} \tag{3}$$

mAP and AP are metrics used to evaluate multi-class classification problems. mAP is the average of the AP values across all classes, while AP is calculated separately for each class. The formulas are shown in Equations (4) and (5).

$$AP = \int_0^1 P(R)\, dR \tag{4}$$

$$mAP = \frac{\sum_{j=1}^{S} AP(j)}{S} \tag{5}$$

In the above formula, S represents the total number of classes, TP is True Positives, FP is False Positives, and FN is False Negatives. $P(R)$ is the precision–recall curve. These metrics are important for evaluating the performance and adaptability of the model.

The False Negative Rate (FNR) is defined as the proportion of positive samples that the model fails to detect. It reflects the model's missed detection rate, as shown in Equation (6).

$$FNR = \frac{FN}{TP + FN} = 1 - \text{Recall} \tag{6}$$

In traditional classification tasks, the False Positive Rate (FPR), defined as the proportion of non-defective samples incorrectly classified as defective, is commonly used to evaluate model performance. However, in the context of industrial defect inspection, this metric may not provide a meaningful assessment of model reliability. To more effectively capture the practical impact of false alarms in such scenarios, this study adopts the False Discovery Rate (FDR) as a key evaluation metric. FDR is defined as the proportion of false positives among all samples predicted as defective, thereby directly quantifying the severity of the model's false alarms. The formulation of FDR is presented in Equation (7).

$$FDR = \frac{FP}{TP + FP} = 1 - \text{Precision} \tag{7}$$

5.4. Results

5.4.1. Training Visualization

Through the experiments described above, we obtained the ES-YOLO framework's experimental results on the N-neudet training set, as illustrated in Figure 5.

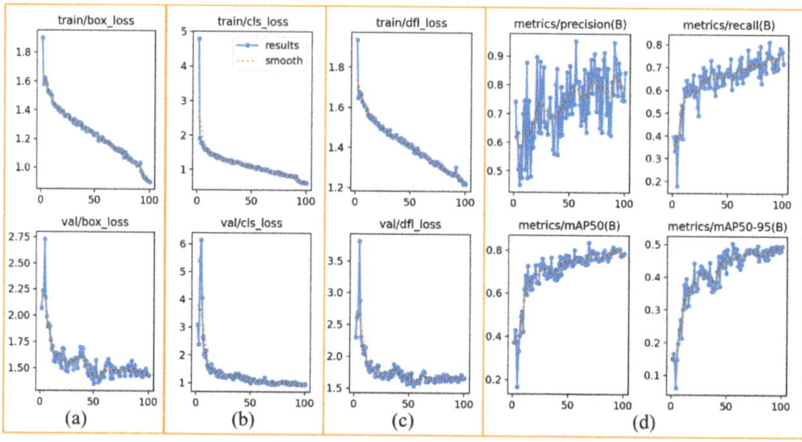

Figure 5. Training result figures. (**a**) The bounding box loss (box_loss), which measures the error in predicted object locations, shows a consistent downward trend, indicating that the model is effectively learning to locate objects; (**b**) the classification loss (cls_loss), indicating the error in predicting the correct object class, also steadily decreases as training progresses; (**c**) Distribution Focal Loss (dfl_loss), another component for bounding box regression, follows a similar convergence pattern; (**d**) Key evaluation metrics, showing the progression of precision, recall, and mean Average Precision (mAP), all exhibit a clear upward trend, signifying significant performance improvement over time.

In Figure 5a, the train/box_loss exhibited a steady downward trend, indicating that the model's bounding box regression capability was gradually improved. The val/box_loss also showed a similar downward trend, confirming the model's strong generalization ability on unseen data. In Figure 5b, both train/cls_loss and val/cls_loss decreased significantly, signifying that the model was effectively learning to distinguish between different object classes. The distribution focal loss, another component of bounding box regression, was shown in Figure 5c. Its convergence on both sets further confirmed the stability of the training process.

In Figure 5d, as the number of training epochs increased, the model's precision gradually improved. The mAP@0.5 (mean average precision at an IoU threshold of 0.5) increased rapidly and stabilized, indicating a significant enhancement in the model's overall detection performance. Although the mAP@0.5:0.95 (mean average precision across IoU thresholds from 0.5 to 0.95) was lower than mAP@0.5, its upward trend was also evident, further demonstrating the model's robustness across different IoU thresholds.

Overall, the ES-YOLO framework exhibited good convergence during training, with both training and validation losses showing a downward trend. Additionally, the model's precision and recall demonstrated significant improvement, highlighting the model's effectiveness in the defect detection task.

5.4.2. Validating Visualization

We selected a subset of images from the N-neudet event dataset as the validation set. After 100 epochs of training, the model's performance on the validation set is presented in Table 2.

Table 2. ES-YOLO validating results.

Class	P	R	mAP@0.5	FNR	FDR
All	0.952	0.657	0.801	0.343	0.048
Crazing	0.817	0.0714	0.238	0.9286	0.183
Inclusion	0.957	1	0.995	0	0.043
Patches	0.982	0.75	0.971	0.25	0.018
Pitted	1	0.62	0.876	0.38	0
Rolled	0.956	0.5	0.732	0.5	0.044
Scratches	0.998	1	0.995	0	0.002

The results showed that the model achieved a precision of 0.952, a recall of 0.657, and an mAP@0.5 of 0.801 across all six categories, demonstrating good overall performance. However, for the two defect types, Crazing and Rolled-in-scale, although the detection precision exceeded 0.8, the recall rates were notably low. This phenomenon indicated that the model struggled to accurately distinguish between these two specific defect types, often leading to misclassifications. These findings highlighted certain limitations of event cameras in fine-grained defect classification tasks.

The qualitative results of the validation set are illustrated in Figure 6. The bounding boxes in Figure 6a represent the ground truth. Figure 6b illustrates the prediction results of the validation set.

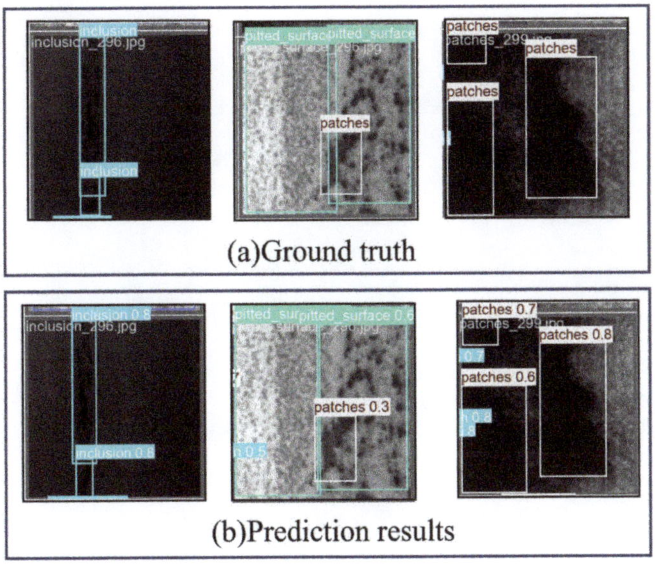

Figure 6. Visualization of validation set results. (**a**) represents the ground truth; (**b**) represents the prediction results of the validation set.

Meanwhile, the original NEU-DET dataset was trained using the YOLOv8s model, and detection results were obtained on the validation set. Table 3 presents a quantitative comparison of the detection performance between models trained on the original NEU-DET RGB dataset and the synthesized N-NEUDET event dataset. The evaluation was conducted using the four key metrics of precision (P), recall (R), mAP@0.5, and inference time. The results demonstrated that, on the validation set, the ES-YOLO model achieved substantially higher detection accuracy on the event-based dataset compared to the RGB dataset, which validated its effectiveness in defect identification.

Table 3. Comparison of training results between RGB and event dataset.

Framework	Dataset	P	R	mAP@0.5	Inference Time
ES-YOLO	N-neudet	0.952	0.657	0.801	79.167 ms
YOLO	NEU-DET	0.84	0.637	0.773	69.444 ms

6. Performance Evaluation of ES-YOLO Under Different Illuminations

To validate the advantages of event cameras in defect detection, we conducted comparative experiments. Specifically, we used the Intel RealSense L515 RGB camera and the DVXplorer Mini event camera to collect data from steel and acrylic pipelines. The trained ES-YOLO framework was then applied for detection, and the results were analyzed for comparison.

The geometric specifications and defect characteristics of the two pipeline types are depicted in Figure 7. Figure 7a presents the steel pipeline, which was sourced from a decommissioned industrial system and contained naturally occurring defects such as rust patches, surface scratches, and pitting corrosion. In contrast, Figure 7b displays the acrylic pipe, on which scratches were artificially created to simulate common damage, given its material properties. In this experiment, we used the detection rate (D_r) metric to evaluate the performance of the ES-YOLO framework. The detection rate was defined as the ratio of the number of defect samples successfully detected by the framework to the total number of defect samples captured by the camera, as expressed in Equation (8).

$$D_r = \frac{D_s}{D_s + D_f} \tag{8}$$

where D_s represents the number of defect samples successfully detected, and D_f represents the number of defect samples not detected.

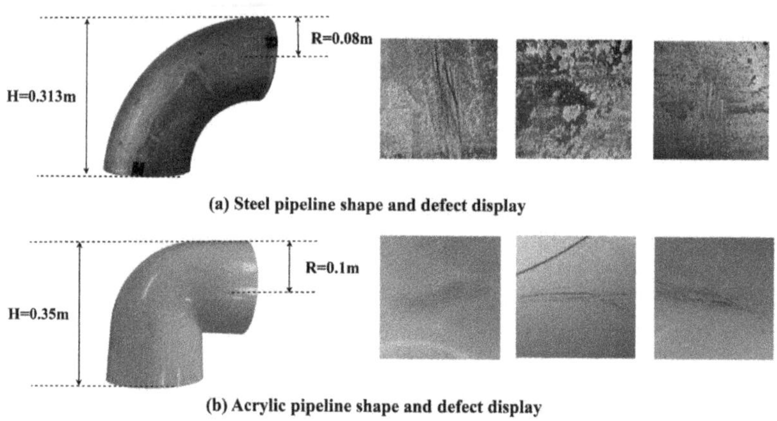

Figure 7. Diagram showing the radius height of the experimental pipeline and its defects.

In the actual pipeline inspection experiment, we designed six comparative experiments to evaluate the detection performance of event cameras and RGB cameras under the same target defects and lighting conditions. First, three target defects were selected on both steel and acrylic pipelines, with their specific locations marked as A, B, and C in Figures 8 and 9. Then, we used an optical illuminance meter UT383 to measure the brightness of the target points, where the illuminance of defects A, B, and C was 45 Lux, 13 Lux, and 2 Lux, respectively. Subsequently, both event and RGB cameras were used to capture images of the target defects on both types of pipelines. A total of 50 images were collected for each target defect using the Intel RealSense L515 RGB camera and the DVXplorer Mini event-based camera, respectively. The collected images were then fed into the ES-YOLO framework for detection.

Figures 8 and 9 present the defect images collected from the steel and acrylic pipelines, respectively. To facilitate a direct comparison of the performance between the two sensors, these figures also include the corresponding detection results obtained from the RGB camera and the event camera under three distinct lighting conditions. Notably, in the low-light environment of 2 Lux, the RGB camera exhibited markedly inferior detection performance compared to the event camera, with frequent occurrences of misclassifications and even complete missed detections. The quantitative detection results for the steel and acrylic pipelines are detailed in Table 4 and Table 5, respectively.

Table 4. Comparison results of event camera and RGB camera detection in steel pipes.

Target Defect	A	B	C
Illuminance	45 Lux	13 Lux	2 Lux
Event camera D_r	94%	90%	84%
RGB camera D_r	66%	30%	10%

Table 5. Comparison results of event camera and RGB camera detection in acrylic pipes.

Target Defect	A	B	C
Illuminance	45 Lux	13 Lux	2 Lux
Event camera D_r	82%	76%	68%
RGB camera D_r	34%	10%	0%

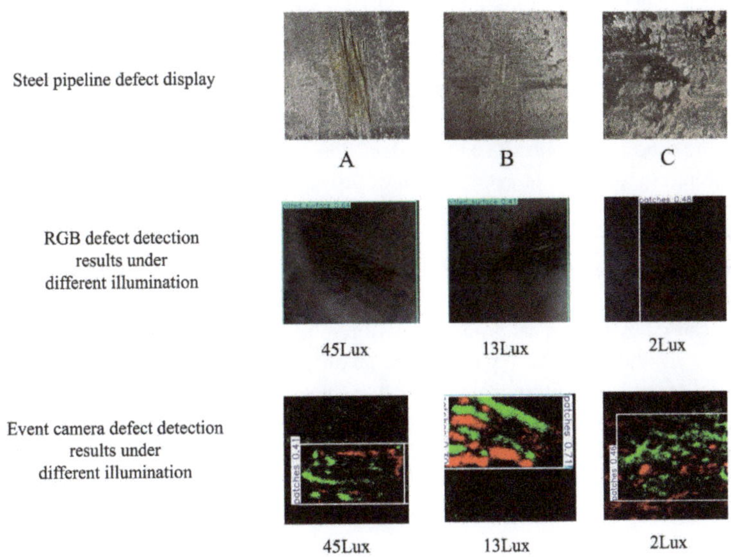

Figure 8. Comparison experiment of different defects and illumination levels in steel pipes. Among them, A, B, and C represent the target defects in the steel pipe. Both the event camera and RGB camera can detect the defect at all three points (A, B, and C), where the illumination levels are 45 Lux, 13 Lux, and 2 Lux, respectively.

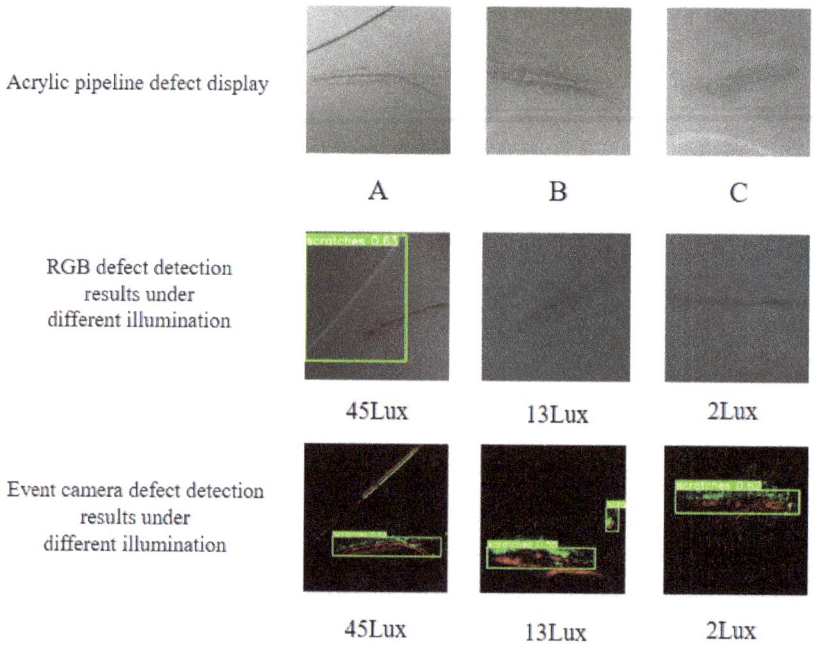

Figure 9. Comparison experiment of different defects and illumination levels in acrylic pipes. Among them, A, B, and C represent the target defects in the acrylic pipe. The illumination levels are 13 Lux and 2 Lux when measured at points B and C, respectively. As a result, only the event camera successfully detected the defects.

The experimental results showed that in the steel pipeline, the event camera successfully detected all target defects, achieving a detection rate of over 84%. In contrast, the defect detection rate using RGB images under 2 Lux low-light conditions was only 10%, which was significantly lower than the detection performance achieved with event-based data. In the acrylic pipeline, the RGB camera detected defects only under 45 Lux and 13 Lux lighting conditions, with a maximum detection rate of 34%. In contrast, the event camera accurately detected the target defects under all lighting conditions, with a detection rate consistently above 60%. These results indicate that, compared to the RGB camera, the event camera could effectively detect pipeline defects in low-light environments. However, the detection rate of the RGB model for acrylic pipelines was significantly lower than that for steel pipelines, primarily due to the traditional RGB models' susceptibility to interference caused by the optical properties of different materials. In contrast, the event camera effectively alleviated the impact of material differences by capturing the dynamic characteristics of defects. The results indicate that the ES-YOLO framework performed better in cross-material detection.

In summary, under the low-light environments, the event camera consistently performed better than the RGB camera, particularly in low-light and cross-material scenarios, which made it more suitable for practical applications and provided a significant overall advantage.

7. Experimental Design for Pipeline Defect Detection via a Snake-like HRM

7.1. Composition of the Pipeline Defect Detection System

To evaluate the performance of the ES-YOLO framework in real-world pipeline environments, a pipeline defect detection experimental platform is constructed, comprising two main components. The first is a visual perception module based on the ES-YOLO framework, designed to tackle detection challenges under low-light conditions. The second is a modeling and control algorithm for the snake-like HRM, developed to enable collision-free navigation within curved pipeline structures. The system consists of an HRM, a sliding rail, a base, an event camera mounted on the end of the HRM, and a PC-based control system, as illustrated in Figure 10.

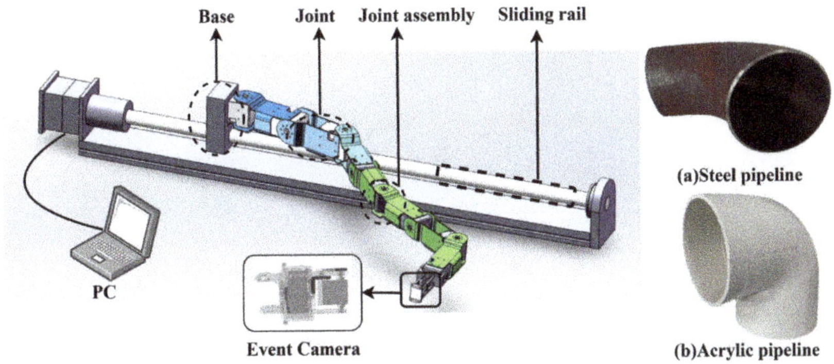

Figure 10. Schematic diagram of the pipeline defect detection system, where (**a**) represents the steel pipeline and (**b**) corresponds to the acrylic pipeline.

A key design consideration is achieving a balance between mechanical strength and maneuverability. The HRM adopts a "3 + 5" hierarchical configuration: three high-torque XH540-W270-R motors connected by magnesium–aluminum alloy joints form the base section, while five lightweight XH430-W270-R motors connected via resin connectors comprise the end section. The sliding rail structure employs a ball screw drive system, which consists primarily of a stepper motor, a lead screw, and a mounting base for the HRM. The ball screw system offers a travel distance of 1.1 m and features high load capacity, and stable stepping performance. The PC-based control system independently communicates control commands to both the HRM and the sliding rail using the RS485 protocol, ensuring reliable and synchronized motion control. Additionally, the DVXplorer Mini event camera is mounted on the HRM's end-effector to capture pipeline defects throughout the insertion process. Table 6 shows more details of the system specifications.

Table 6. Composition of the pipeline defect detection system.

Component	Specifications
HRM	Three high-torque XH540-W270-R motors
	Five lightweight XH430-W270-R motors
The sliding rail	Stepper motor
	Lead screw
	Base
Controller	PC(Intel(R) Core(TM) i5-8265U CPU @ 1.60GHz)
Protocol	RS485 protocol to control HRM
	RS485 protocol to control sliding rail
Event camera	DVXplorer Mini

The overall workflow of the system is illustrated in Figure 11. First, a control algorithm facilitates its collision-free insertion to accurately reach the target detection position.

Subsequently, the event camera at the end of the HRM is used to collect internal data from within the pipeline. Finally, the acquired data is processed using the ES-YOLO framework, and defect detection results are produced.

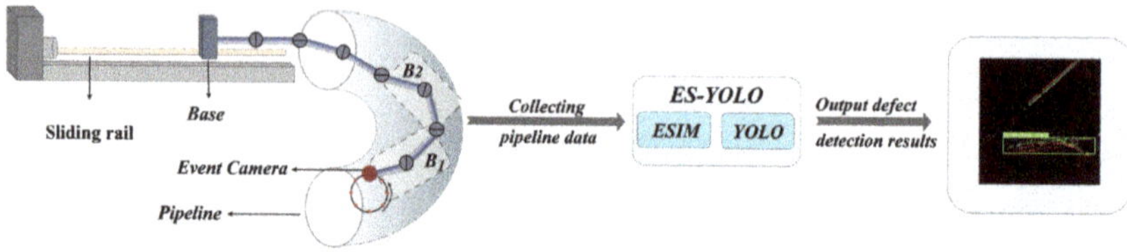

Figure 11. Schematic diagram of the pipeline defect detection experimental workflow.

7.2. The Model and Control Algorithm of the HRM

To enable collision-free defect inspection within narrow pipelines, this study employs a motor-driven, 9-degree-of-freedom (DOF) hyper-redundant snake-like robotic arm. The HRM consists of a base, a sliding rail, eight joints, and joint assemblies, with its detailed configuration illustrated in Figure 12.

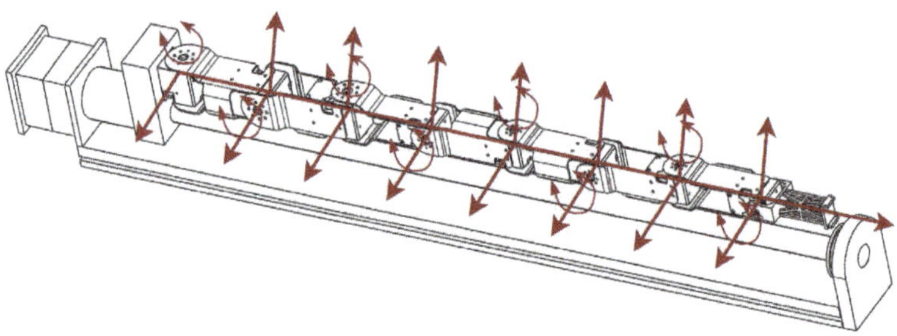

Figure 12. Mechanical structure design of HRM and sliding rail.

To characterize the motion properties of the HRM, this article establishes its kinematic model based on forward kinematics and Denavit–Hartenberg (D-H) parameters. The relevant D-H parameters are presented in Table 7. Based on this model, the homogeneous transformation matrix ${}^{i}_{i-1}T(i = 1,2\ldots n)$ between adjacent joints is derived, as shown in Equation (9), where 's' represents the sine function (sin), and 'c' represents the cosine function (cos).

$$
{}^{i}_{i-1}\boldsymbol{T} = \begin{bmatrix} c\theta_i & -s\theta_i c\alpha_i & s\theta_i s\alpha_i & a_i c\theta_i \\ s\theta_i & c\theta_i c\alpha_i & -c\theta_i s\alpha_i & a_i s\theta_i \\ 0 & s\alpha_i & c\alpha_i & d_i \\ 0 & 0 & 0 & 1 \end{bmatrix} \tag{9}
$$

By combining the transformation matrix ${}^{0}_{w}\boldsymbol{T}$ from the world coordinate system to the base coordinate system and the homogeneous transformation matrix ${}^{n}_{w}\boldsymbol{T}$ from the world coordinate system to the end-effector, the forward kinematics equation is obtained, as shown in Equation (10). Here, $\boldsymbol{q} = [\theta_1, \theta_2, \ldots, \theta_n]$ denotes the configuration vector of the HRM.

$$
\text{fkine}(\boldsymbol{q}) = {}^{n}_{w}\boldsymbol{T} = {}^{0}_{w}\boldsymbol{T}\,{}^{1}_{0}\boldsymbol{T}\cdots{}^{n}_{n-1}\boldsymbol{T} \tag{10}
$$

Due to the high degree of redundancy in the HRM and the spatial constraints of narrow pipeline environments, the control complexity of pipeline defect inspection tasks is significantly increased. To ensure collision-free operation of the HRM during inspection, efficient path planning is essential. However, due to kinematic constraints and the inherent complexity of the planning process, finding an optimal path often incurs substantial time and computational costs, and requires precise control of each HRM joint to avoid collisions. To balance planning efficiency and path quality, this article adopts an efficient strategy that seeks a suboptimal HRM configuration to reduce computational cost. This strategy is ultimately implemented by combining Nonlinear Model Predictive Control (NMPC) with a Snake-inspired Crawling Algorithm (SCA) [12]. The obstacle avoidance strategy for pipeline defect inspection includes two stages: insertion and inspection, as shown in Figure 13. Experimental validation demonstrates that the minimum pipeline the HRM can successfully traverse is a 90° curved pipe with an inner diameter of 8 cm and a length of 48 cm. This limit is jointly determined by the physical dimensions of the manipulator, the pipeline radius, and the joint bending capability.

Table 7. D-H parameters of the snake manipulator.

Link i	$\theta_i\,(^\circ)$	$\alpha_i\,(^\circ)$	$a_i\,(m)$	$d_i\,(m)$
$i = 1,3,5,7$	θ_i	-90°	0.1	0
$i = 2,4,6,8$	θ_i	90°	0.1	0

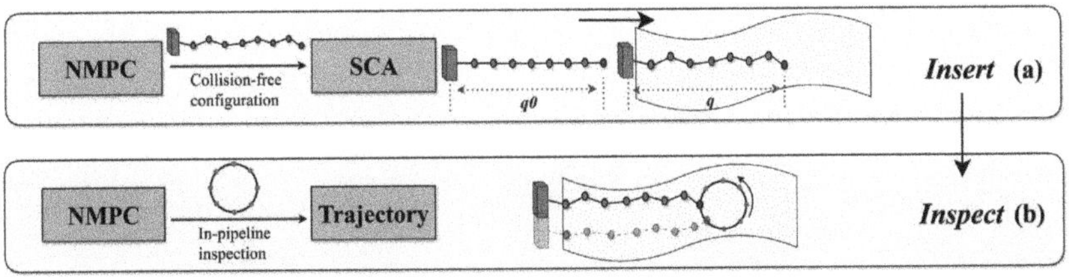

Figure 13. Flow chart of the pipeline inspection process: q_0 (initial horizontal configuration), q (collision-free configuration).

7.3. Generalization Ability of NMPC

To achieve comprehensive data acquisition of the inner pipeline wall, the event camera mounted on the HRM's end-effector performs a rotational scan along the pipeline cross-section. This scanning trajectory is discretized into $M_i(i = 1, \ldots, m)$ key target points. To ensure the HRM can safely and accurately reach each target point, the NMPC algorithm is employed to pre-plan a corresponding collision-free configuration, $\boldsymbol{q}_i \in \mathbb{R}^n$, for every target point before the manipulator enters the pipeline.

The optimization objective of this pose planning is to minimize the sum of the absolute values of the joint angles and the sum of the changes in joint angles between adjacent configurations. This objective helps prevent the HRM from moving to its joint limits and avoids abrupt movements caused by local minima. The planning method for the collision-free configuration of the i-th target point is shown in Equation (11), where θ_{ij} represents the j-th joint angle ($j = 1, \ldots, n$) of the i-th configuration; x_i is the desired pose vector of the i-th target point; θ_{lb}, θ_{ub} define the upper and lower limits of the HRM's joint angles; and D_{safe} represents the minimum safe distance between the HRM and the pipeline.

$$\begin{aligned} &\text{minimize} \sum_{j=1}^{n} |\theta_{ij}| + \sum_{j=1}^{n} |\theta_{ij} - \theta_{i-1j}| \\ &\text{subject to} \begin{cases} \|x_i - \text{pose}(\text{fkine}(\boldsymbol{q}_i))\|_2 = 0 \\ D^j_{\text{pipe}} \leqslant R - D_{\text{safe}} \\ \theta_{ij} \in [\theta_{\text{lb}}, \theta_{\text{ub}}] \\ \Delta\theta_{\text{lb}} \leqslant |\theta_{ij} - \theta_{i-1j}| \leqslant \Delta\theta_{\text{ub}} \end{cases} \end{aligned} \tag{11}$$

A solution from the NMPC is not always guaranteed due to multiple constraints, including the desired pose of the HRM end-effector, pipeline curvature, radius, and the safe distance D_{safe}. In the experiments, D_{safe} is set to 4 cm. This threshold accounts for both the physical radius of the HRM links and potential positioning errors during actual operation, thereby ensuring physical safety. Planning is considered to have failed under two conditions.

To systematically evaluate the generalization ability of NMPC, this article conducts a series of collision-free configuration simulations in MATLAB. The objective of NMPC is to compute safe, collision-free configurations. For the HRM, configurations closer to the pipeline centerline reduce the risk of collision with the wall and thereby improve safety. Therefore, this simulation aims to calculate the maximum distance, D_{max}, that the HRM's configuration deviates from the centerline as it passes through pipelines of different curvatures. The minimum pipeline radius for ensuring safe passage is then calculated by adding this maximum deviation distance, D_{max}, to a predefined safety threshold, D_{safe}, where curvature is defined as the included angle of the sector forming the bent pipe section. The experiments test three typical pipeline curvatures, including 45°, 90°, 135°. The total length of all simulated pipelines is kept consistent, and their shapes are illustrated in Figure 14. Although the simulation results cannot directly quantify the algorithm's generalization ability, they provide a qualitative evaluation of its adaptability under different geometric constraints. The specific experimental results are detailed in Table 8.

Table 8. Minimum safe passage radius of the HRM for different pipeline curvatures.

Curvatures	D_{safe}	D_{max}	Minimum Pipeline Radius for Safe Passage
45°	4 cm	2.53 cm	6.53 cm
90°	4 cm	3.72 cm	7.72 cm
135°	4 cm	5.13 cm	9.13 cm

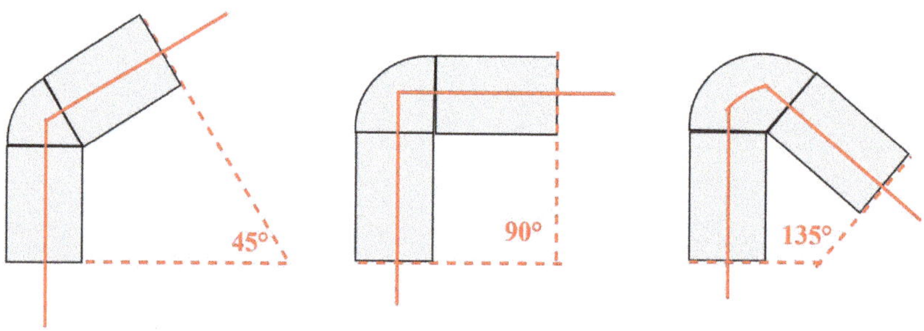

Figure 14. Schematic diagram of single-bend pipe models with different curvatures, where the red solid line represents the pipeline centerline.

The experimental results indicate that when the NMPC algorithm generates a collision-free configuration for a specific target point, higher pipeline curvature requires a larger radius to ensure safe passage. Specifically, for a pipeline with a 45° curvature, the minimum passage radius required by the collision-free configuration planned by NMPC is 6.53 cm. When the curvature is 90°, the minimum pipeline radius the HRM can pass through is 7.72 cm. For a high-curvature pipeline of 135° , the pipeline radius must be at least 9.13 cm to ensure the HRM can perform the inspection safely. Figure 15 illustrates a successful planning result for a pipeline with a radius of 10 cm and a curvature of 90°, along with the corresponding collision-free configuration of the HRM.

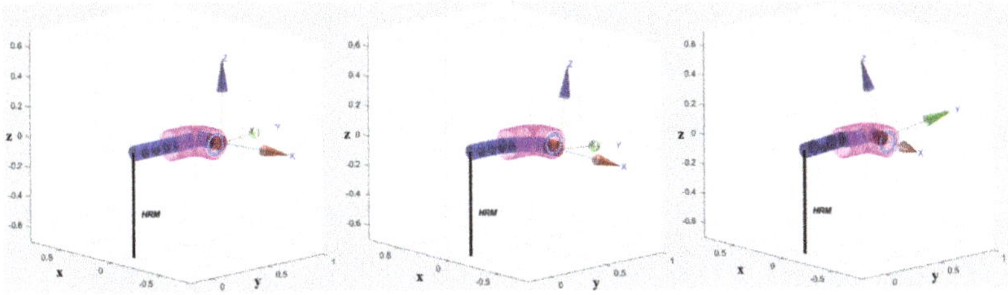

Figure 15. Simulation results for the collision-free inspection of the pipeline defect detection system.

8. Experimental Results for Pipeline Defect Detection via a Snake-like HRM

To validate the practicality and feasibility of the proposed pipeline defect detection system, physical experiments were conducted on both steel and acrylic pipelines, as illustrated in Figure 16. In each pipeline environment, the NMPC optimization algorithm was employed to compute collision-free configurations and generate corresponding motion trajectories.

Figure 16. Experimental platform for pipeline defect detection based on an event camera. The setup comprises the key components of the system, including a HRM, a movable base, a sliding rail, an event-based camera, and the pipeline specimen under inspection.

The generation of the HRM's collision-free configurations was conducted through MATLAB simulations. Detailed numerical results are presented in Table 9. The table reports the maximum positioning error between the end-effector and the target defect. These metrics are provided for both steel and acrylic pipeline environments. To visually demonstrate the insertion process of the HRM, six consecutive visualization frames are presented in Figure 17. During the pipeline insertion process, the illumination from the pipe entrance to the darkest section ranged between 10 Lux and 50 Lux, corresponding to a low-light environment. Owing to its high dynamic range and asynchronous sensing mechanism, the event camera effectively responded to relative changes in brightness and continuously output an event stream, thereby capturing defect details clearly from the bright pipe entrance to the dim depths of the pipeline. Subsequently, the event camera was inserted into the pipeline by the manipulator to collect data, and the detection results were obtained through ES-YOLO analysis.

Table 9. Geometries of the two pipelines and experimental error results.

Pipe Type	Radius	Pipe Angle	Maximum Error
Steel	8 cm	90°	1.542 cm
Acrylic	10 cm	90°	1.947 cm

Based on the aforementioned experiments, the detection results of the proposed pipeline defect detection system are presented in Figures 18 and 19. The results indicate that the target defects were successfully detected by inputting the pipeline data, collected by the event camera mounted on the HRM, into the ES-YOLO framework. These results demonstrate that the bio-inspired vision-based system could effectively identify defects in pipelines made of different materials.

Figure 17. The process of defect detection inside the pipeline using an event camera mounted on the HRM. Six consecutive frames illustrate the collision-free insertion of the HRM into the pipeline. The illuminance at the pipeline entrance is approximately 50 Lux. As the HRM advances deeper into the pipeline, occlusion of the entrance light source by its arm and its own shadow causes the light to diminish, with the minimum illuminance inside the pipeline dropping to approximately 10 Lux.

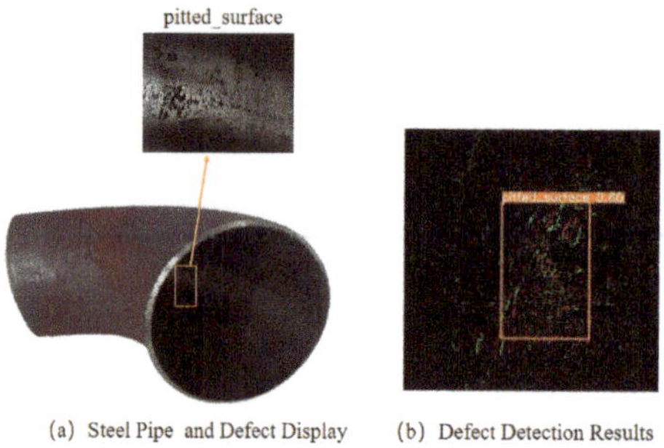

Figure 18. Steel pipe internal defect detection. (**a**) shows the shape and defects of the steel pipe; (**b**) shows the detection results of the steel pipe.

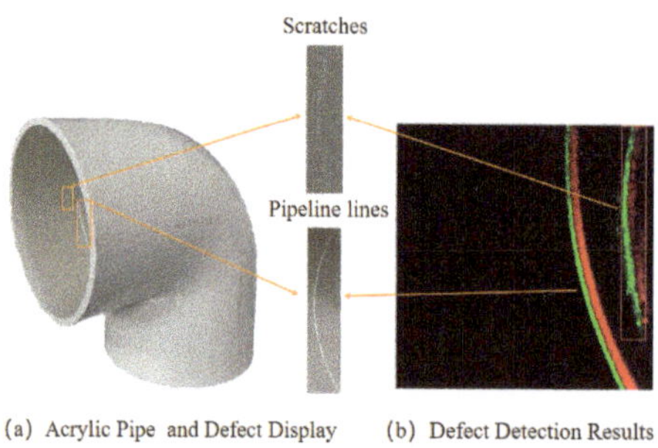

Figure 19. Acrylic pipe internal defect detection. (**a**) shows the structure and defects of the acrylic pipe; (**b**) shows the detection results of the acrylic pipe defects.

In summary, the experimental results demonstrate that the ES-YOLO framework exhibited strong applicability in practical pipeline defect detection and possessed cross-material generalization capability, which validates the robustness and effectiveness of the proposed system.

9. Discussion

In this study, the ES-YOLO framework successfully detects defects in both steel and acrylic pipes, demonstrating that generating event data through simulation is an efficient and feasible strategy to address dataset scarcity in specialized application domains. This finding establishes a reproducible, data-driven paradigm that can be extended to other industrial and scientific fields lacking native event datasets.

Comparative experiments with traditional RGB vision further highlight the advantages of event cameras in low-light conditions. These results not only confirm their technical feasibility but also open new avenues for robotic applications in visually degraded environments such as unlit pipelines, nighttime inspections, and settings with high-dynamic-range illumination changes. Notably, the system achieves defect detection in near-total darkness, a task that remains challenging for conventional RGB-based methods.

Moreover, the strong robustness exhibited across both steel and acrylic pipelines warrants deeper investigation. We speculate that this cross-material generalization arises from the intrinsic properties of the RGB-to-event conversion process, which primarily encodes brightness changes rather than absolute color or texture information. By filtering out material-dependent surface details while emphasizing geometric contours of defects, this process enhances detection consistency across diverse environments—a property of considerable practical significance for real-world inspection systems.

In this study, the HRM exhibits notable advantages in confined and complex environments that demand high precision and stability. It enables stable, high-precision, non-contact data acquisition and offers a greater payload capacity for carrying advanced sensor suites. Additionally, its energy consumption is significantly lower than that of UAVs. However, the HRM is less suitable for large-scale environments, such as wide-diameter pipelines or oil storage tanks, where UAVs and other mobile robots demonstrate clear advantages. Notably, the ES-YOLO framework is highly portable and can be deployed across various robotic platforms, including UAVs and ground-based mobile robots.

Ultimately, the HRM, UAV, or other robotic systems serve merely as carriers for executing pipeline inspection tasks. Deploying the ES-YOLO framework on different platforms to support diverse engineering applications presents broad prospects for practical deployment.

However, the current approach has a fundamental limitation: to leverage the mature YOLOv8 architecture, the asynchronous event stream must first be converted into discrete event frames before being input into the ES-YOLO framework. While this conversion step facilitates the application of advanced detection algorithms, it compromises the intrinsic advantages of event cameras, such as microsecond-level temporal resolution and low latency. Therefore, future research should prioritize overcoming this bottleneck by exploring advanced algorithms capable of directly processing raw event streams. A particularly promising direction is the use of Spiking Neural Networks (SNNs), whose event-driven and asynchronous processing characteristics are inherently compatible with the data format of event cameras, offering the potential for low-latency and low-power defect detection. In addition, asynchronous convolutional networks and graph-based methods that operate directly on event streams also warrant further investigation. Ultimately, the goal is to deeply integrate these emerging technologies with the HRM detection system to enhance the generalizability and practical utility of the entire framework.

10. Conclusions

This article proposed an ES-YOLO framework to enhance defect detection in low-light pipeline environments, which leveraged an event camera based on bionic visual perception. To address the issue of scarce event datasets, the RGB NEU-DET dataset was converted into the N-neudet event-based dataset using the ESIM module. Subsequently, the pipeline defect detection model was trained on the generated event data via YOLOv8s. To verify the accuracy of this framework in low-light environments, two sets of comparative experiments were conducted to assess the defect detection performance of RGB cameras versus event cameras within both steel and acrylic pipelines. The results showed that RGB cameras fail to effectively identify defects, while event cameras demonstrated clear advantages under low-light conditions. Furthermore, the pipeline defect detection system based on a snake-like HRM was developed, and its feasibility and detection accuracy were validated through physical experiments in realistic pipeline inspection scenarios.

Future work will focus on expanding the breadth and depth of application for the proposed HRM-based detection method. To achieve real-time and efficient system operation, low-power models such as Spiking Neural Networks (SNNs) will be deployed on embedded hardware to reduce inference time and improve energy efficiency. At the system architecture level, migration to the ROS 2 framework will be explored to enhance modularity, stability, and scalability. On the perception level, efforts will target key challenges such as real-time 3D modeling and dynamic path planning in unknown environments—both critical to improving system autonomy. To validate and improve the framework's generalizability, its application will be extended beyond pipeline interiors to encompass a broader range of industrial geometries, including ventilation ducts, mining tunnels, and confined industrial cavities. The ultimate goal is to develop an automated and intelligent inspection system capable of adapting to diverse environments while maintaining high autonomy, efficiency, and robustness for inspection tasks in various constrained spaces.

Author Contributions: Conceptualization, X.X.; methodology, X.X. and M.S.; software, M.S. and B.G.; validation, M.S., B.G. and J.W. (Jingxue Wu); data curation, B.G.; writing—original draft preparation, M.S.; writing—review and editing, X.X., J.W. (Jianming Wang) and J.L. All authors have read and agreed to the published version of the manuscript.

Funding: This research received no external funding.

Institutional Review Board Statement: Not applicable

Informed Consent Statement: Not applicable

Data Availability Statement: The original contributions presented in this study are included in the article.

Conflicts of Interest: The authors declare no conflicts of interest.

Abbreviations

The following abbreviations are used in this manuscript:

HRM	Hyper-Redundant Manipulator
NMPC	Nonlinear Model Predictive Control
SCA	Snake-inspired Crawling Algorithm
DOF	Degrees Of Freedom
D-H	Denavit–Hartenberg

References

1. Kazeminasab, S.; Sadeghi, N.; Janfaza, V.; Razavi, M.; Ziyadidegan, S.; Banks, M.K. Localization, mapping, navigation, and inspection methods in in-pipe robots: A review. *IEEE Access* **2021**, *9*, 162035–162058. [CrossRef]
2. Le, D.V.K.; Chen, Z.; Rajkumar, R. Multi-sensors in-line inspection robot for pipe flaws detection. *IET Sci. Meas. Technol.* **2020**, *14*, 71–82. [CrossRef]
3. Gehrig, D.; Gehrig, M.; Hidalgo-Carrió, J.; Scaramuzza, D. Video to events: Recycling video datasets for event cameras. In Proceedings of the IEEE/CVF Conference on Computer Vision and Pattern Recognition, Seattle, WA, USA, 13–19 June 2020; pp. 3586–3595.

4. Rebecq, H.; Gehrig, D.; Scaramuzza, D. ESIM: An open event camera simulator. In Proceedings of the Conference on Robot Learning, PMLR, Stockholm, Sweden, 10–15 July 2018; pp. 969–982.
5. Shao, L.; Wang, Y.; Guo, B.; Chen, X. A review over state of the art of in-pipe robot. In Proceedings of the 2015 IEEE International Conference on Mechatronics and Automation (ICMA), Beijing, China, 2–5 August 2015; pp. 2180–2185.
6. Kazeminasab, S.; Aghashahi, M.; Banks, M.K. Development of an inline robot for water quality monitoring. In Proceedings of the 2020 5th International Conference on Robotics and Automation Engineering (ICRAE), Singapore, 20–22 November 2020; pp. 106–113.
7. Liljebäck, P.; Pettersen, K.Y.; Stavdahl, Ø.; Gravdahl, J.T. A review on modelling, implementation, and control of snake robots. *Robot. Auton. Syst.* **2012**, *60*, 29–40. [CrossRef]
8. Ji, H.; Xie, H.; Yang, H. A Spatial Path Following Method for Hyper-Redundant Manipulators by Step-by-Step Search and Calculating. In Proceedings of the 2022 7th International Conference on Robotics and Automation Engineering (ICRAE), Singapore, 18–20 November 2022; pp. 292–298. [CrossRef]
9. Soltani, A.; Tawfik, H.; Goulermas, J.; Fernando, T. Path planning in construction sites: Performance evaluation of the Dijkstra, A*, and GA search algorithms. *Adv. Eng. Inform.* **2002**, *16*, 291–303. [CrossRef]
10. Ji, H.; Xie, H.; Wang, C.; Yang, H. E-RRT*: Path Planning for Hyper-Redundant Manipulators. *IEEE Robot. Autom. Lett.* **2023**, *8*, 8128–8135. [CrossRef]
11. Liu, J.; Tong, Y.; Liu, J. Review of snake robots in constrained environments. *Robot. Auton. Syst.* **2021**, *141*, 103785. [CrossRef]
12. Zhu, J.; Su, M.; Li, L.; Xiang, Y.; Wang, J.; Xiao, X. Snake-inspired trajectory planning and control for confined pipeline inspection with hyper-redundant manipulators. *Biomim. Intell. Robot.* **2025**, *5*, 100245. [CrossRef]
13. Liu, T.; Jiang, Z.P. Event-based control of nonlinear systems with partial state and output feedback. *Automatica* **2015**, *53*, 10–22. [CrossRef]
14. Liu, T.; Qin, Z.; Hong, Y.; Jiang, Z.P. Distributed Optimization of Nonlinear Multiagent Systems: A Small-Gain Approach. *IEEE Trans. Autom. Control* **2022**, *67*, 676–691. [CrossRef]
15. Jin, Z. Global Asymptotic Stability Analysis for Autonomous Optimization. *IEEE Trans. Autom. Control* **2025**, 1–7. [CrossRef]
16. Xu, C.; Liu, Z.; Hu, C.; Li, X. Improved Hybrid A* Algorithm Obstacle Avoidance Strategy Based on Reinforcement Learning. In Proceedings of the 2023 42nd Chinese Control Conference (CCC), Tianjin, China, 24–26 July 2023; pp. 4077–4082. [CrossRef]
17. Song, R.; Liu, Y.; Bucknall, R. Smoothed A* algorithm for practical unmanned surface vehicle path planning. *Appl. Ocean Res.* **2019**, *83*, 9–20. [CrossRef]
18. Yu, D.; Roh, M.I. Method for anti-collision path planning using velocity obstacle and A* algorithms for maritime autonomous surface ship. *Int. J. Nav. Archit. Ocean. Eng.* **2024**, *16*, 100586. [CrossRef]
19. Tang, X.; Zhou, H.; Xu, T. Obstacle avoidance path planning of 6-DOF robotic arm based on improved A* algorithm and artificial potential field method. *Robotica* **2024**, *42*, 457–481. [CrossRef]
20. Luo, M.; Hou, X.; Yang, J. Surface Optimal Path Planning Using an Extended Dijkstra Algorithm. *IEEE Access* **2020**, *8*, 147827–147838. [CrossRef]
21. Orchard, G.; Jayawant, A.; Cohen, G.K.; Thakor, N. Converting static image datasets to spiking neuromorphic datasets using saccades. *Front. Neurosci.* **2015**, *9*, 437. [CrossRef]
22. Amir, A.; Taba, B.; Berg, D.; Melano, T.; McKinstry, J.; Di Nolfo, C.; Nayak, T.; Andreopoulos, A.; Garreau, G.; Mendoza, M.; et al. A low power, fully event-based gesture recognition system. In Proceedings of the IEEE Conference on Computer Vision and Pattern Recognition, Honolulu, HI, USA, 21–26 July 2017; pp. 7243–7252.
23. Calabrese, E.; Taverni, G.; Awai Easthope, C.; Skriabine, S.; Corradi, F.; Longinotti, L.; Eng, K.; Delbruck, T. DHP19: Dynamic vision sensor 3D human pose dataset. In Proceedings of the IEEE/CVF Conference on Computer Vision and Pattern Recognition Workshops, Long Beach, CA, USA, 16–17 June 2019.
24. Binas, J.; Neil, D.; Liu, S.C.; Delbruck, T. DDD17: End-to-end DAVIS driving dataset. *arXiv* **2017**, arXiv:1711.01458.
25. Gehrig, M.; Aarents, W.; Gehrig, D.; Scaramuzza, D. DSEC: A Stereo Event Camera Dataset for Driving Scenarios. *IEEE Robot. Autom. Lett.* **2021**, *6*, 4947–4954. [CrossRef]
26. Cannici, M.; Ciccone, M.; Romanoni, A.; Matteucci, M. Asynchronous convolutional networks for object detection in neuromorphic cameras. In Proceedings of the IEEE/CVF Conference on Computer Vision and Pattern Recognition Workshops, Long Beach, CA, USA, 16–17 June 2019.
27. Maqueda, A.I.; Loquercio, A.; Gallego, G.; García, N.; Scaramuzza, D. Event-based vision meets deep learning on steering prediction for self-driving cars. In Proceedings of the IEEE Conference on Computer Vision and Pattern Recognition, Salt Lake City, UT, USA, 18–23 June 2018; pp. 5419–5427.
28. Alonso, I.; Murillo, A.C. EV-SegNet: Semantic segmentation for event-based cameras. In Proceedings of the IEEE/CVF Conference on Computer Vision and Pattern Recognition Workshops, Long Beach, CA, USA, 16–17 June 2019.

29. Rebecq, H.; Ranftl, R.; Koltun, V.; Scaramuzza, D. Events-to-video: Bringing modern computer vision to event cameras. In Proceedings of the IEEE/CVF Conference on Computer Vision and Pattern Recognition, Long Beach, CA, USA, 16–17 June 2019; pp. 3857–3866.
30. Rebecq, H.; Ranftl, R.; Koltun, V.; Scaramuzza, D. High Speed and High Dynamic Range Video with an Event Camera. *IEEE Trans. Pattern Anal. Mach. Intell.* **2021**, *43*, 1964–1980. [CrossRef]
31. Kaiser, J.; Tieck, J.C.V.; Hubschneider, C.; Wolf, P.; Weber, M.; Hoff, M.; Friedrich, A.; Wojtasik, K.; Roennau, A.; Kohlhaas, R.; et al. Towards a framework for end-to-end control of a simulated vehicle with spiking neural networks. In Proceedings of the 2016 IEEE International Conference on Simulation, Modeling, and Programming for Autonomous Robots (SIMPAR), San Francisco, CA, USA, 13–16 December 2016; pp. 127–134.
32. Li, W.; Saeedi, S.; McCormac, J.; Clark, R.; Tzoumanikas, D.; Ye, Q.; Huang, Y.; Tang, R.; Leutenegger, S. Interiornet: Mega-scale multi-sensor photo-realistic indoor scenes dataset. *arXiv* **2018**, arXiv:1809.00716.
33. Mueggler, E.; Rebecq, H.; Gallego, G.; Delbruck, T.; Scaramuzza, D. The event-camera dataset and simulator: Event-based data for pose estimation, visual odometry, and SLAM. *Int. J. Robot. Res.* **2017**, *36*, 142–149. [CrossRef]
34. Gallego, G.; Lund, J.E.; Mueggler, E.; Rebecq, H.; Delbruck, T.; Scaramuzza, D. Event-based, 6-DOF camera tracking from photometric depth maps. *IEEE Trans. Pattern Anal. Mach. Intell.* **2017**, *40*, 2402–2412. [CrossRef]
35. Gallego, G.; Forster, C.; Mueggler, E.; Scaramuzza, D. Event-based camera pose tracking using a generative event model. *arXiv* **2015**, arXiv:1510.01972. [CrossRef]
36. Jiang, H.; Sun, D.; Jampani, V.; Yang, M.H.; Learned-Miller, E.; Kautz, J. Super slomo: High quality estimation of multiple intermediate frames for video interpolation. In Proceedings of the IEEE Conference on Computer Vision and Pattern Recognition, Salt Lake City, UT, USA, 18–23 June 2018; pp. 9000–9008.
37. Redmon, J.; Divvala, S.; Girshick, R.; Farhadi, A. You only look once: Unified, real-time object detection. In Proceedings of the IEEE Conference on Computer Vision and Pattern Recognition, Las Vegas, NV, USA, 27–30 June 2016; pp. 779–788.
38. Jocher, G.; Chaurasia, A.; Qiu, J. Ultralytics YOLOv8, Version 8.0.0. *Ultralytics*, **2023**. Available online: https://github.com/ultralytics/ultralytics (accessed on 22 August 2025).
39. Bochkovskiy, A.; Wang, C.Y.; Liao, H.Y.M. Yolov4: Optimal speed and accuracy of object detection. *arXiv* **2020**, arXiv:2004.10934. [CrossRef]
40. Khanam, R.; Hussain, M.; Hill, R.; Allen, P. A comprehensive review of convolutional neural networks for defect detection in industrial applications. *IEEE Access* **2024**, *12*, 94250–94295. [CrossRef]
41. Gehrig, D.; Loquercio, A.; Derpanis, K.G.; Scaramuzza, D. End-to-end learning of representations for asynchronous event-based data. In Proceedings of the IEEE/CVF International Conference on Computer Vision, Seoul, Republic of Korea, 27 October–2 November 2019; pp. 5633–5643.
42. Bao, Y.; Song, K.; Liu, J.; Wang, Y.; Yan, Y.; Yu, H.; Li, X. Triplet-graph reasoning network for few-shot metal generic surface defect segmentation. *IEEE Trans. Instrum. Meas.* **2021**, *70*, 1–11. [CrossRef]

biomimetics

Article

Dual-Coated Antireflective Film for Flexible and Robust Multi-Environmental Optoelectronic Applications

Hyuk Jae Jang [1], Jaemin Jeon [1], Joo Ho Yun [2], Iqbal Shudha Tasnim [1], Soyeon Han [2], Heeyoung Lee [2], Sungguk An [2], Seungbeom Kang [3], Dongyeon Kim [3] and Young Min Song [1,4,5,*]

1 School of Electrical Engineering and Computer Science, Gwangju Institute of Science and Technology (GIST), Gwangju 61005, Republic of Korea; hjjang3472@gm.gist.ac.kr (H.J.J.)
2 Samsung Display Co., Ltd., Yongin-si 17113, Republic of Korea
3 School of Mechanical Engineering, Gwangju Institute of Science and Technology (GIST), Gwangju 61005, Republic of Korea
4 Artificial Intelligence (AI) Graduate School, Gwangju Institute of Science and Technology (GIST), Gwangju 61005, Republic of Korea
5 Department of Semiconductor Engineering, Gwangju Institute of Science and Technology (GIST), Gwangju 61005, Republic of Korea
* Correspondence: ymsong@gist.ac.kr

Abstract: Artificial antireflective nanostructured surfaces, inspired by moth eyes, effectively reduce optical losses at interfaces, offering significant advantages in enhancing optical performance in various optoelectronic applications, including solar cells, light-emitting diodes, and cameras. However, their limited flexibility and low surface hardness constrain their broader use. In this study, we introduce a universal antireflective film by integrating nanostructures on both sides of a thin polycarbonate film. One side was thinly coated with Al_2O_3 for its high hardness, enhancing surface durability while maintaining flexibility. The opposite side was coated with SiO_2 to optimize antireflective properties, making the film suitable for diverse environments (i.e., air, water, and adhesives). This dual-coating strategy resulted in a mechanically robust and flexible antireflective film with superior optical properties in various conditions. We demonstrated the universal capabilities of our antireflective film via optical simulations and experiments with the fabricated film in different environments.

Keywords: antireflection; flexible film; robust film; multifunctional film; dual coating; electronic application

1. Introduction

The reduction in optical losses at interfaces between materials with differing refractive indices is critical for optimizing the optical performance of various optoelectronic fields, such as light-emitting diodes, photovoltaic devices, image sensors, cameras, transparent glasses, and energy-harvesting [1–11]. To address this issue, multilayer coatings have traditionally been employed as a common method for producing antireflective surfaces. These coatings are designed to minimize reflection by stacking multiple layers of materials with varying refractive indices [12–14]. However, while effective, multilayer coatings often face challenges related to thermal mismatch, material selection, and the complexity of optical design [14].

In recent years, artificial nanostructured surfaces, inspired by the natural design of moth eyes, cicada wings, and lotus leaves, have been developed with multifunctional nanostructures, such as antireflection, bactericidal properties, and superhydrophobicity [15–18]. Notably, artificial antireflective structures (ARSs) suppress Fresnel reflection at the interface by linearly changing the refractive index [19,20]. These antireflective nanostructured surfaces have emerged as a promising alternative with geometrical optimizations. These nanostructures effectively minimize light reflection, enhance optical performance across a wide range of wavelengths, and have a simple optical design [20]. Despite their advantages, these nanostructured surfaces encounter significant limitations, particularly in terms of flexibility, surface stability, and hardness [21–25]. Their limited mechanical robustness restricts their use in more demanding environments, which hinders the broader application of these antireflective surfaces in advanced optoelectronic technologies [9–11,26–30]. Overcoming these limitations is essential to fully exploit the potential of nanostructured antireflective surfaces.

In this study, we developed a universal antireflective nanostructured polycarbonate film (*uni*-ARS PC film) for flexibility with better robustness by integrating nanostructures on both sides of a thin PC film. Our proposed *uni*-ARS PC film has the advantage of reducing reflectance across various materials with different refractive indices, such as air, water, and adhesive layers, all within a single design. To enhance surface durability without compromising flexibility, one side of the film was thinly coated with aluminum oxide (Al_2O_3), a material known for its high hardness. The opposite side was coated with silicon dioxide (SiO_2), which optimizes antireflective properties, making the film adaptable to different environments, including air, water, and adhesive layers.

This *uni*-ARS PC film, utilizing a dual-coating strategy, resulted in a mechanically robust and flexible antireflective film with superior optical properties across different conditions. To validate our approach, we conducted comprehensive optical simulations and performed measurements on the fabricated film in diverse environments (i.e., air, water, and display). The results demonstrate that our design effectively addresses the mechanical and optical limitations of artificial antireflective surfaces, offering a versatile solution that is suitable for a wide range of applications.

2. Materials and Methods

For the fabrication of the dual-coated antireflective film, a hot-pressing process was initially performed to implement ARS on a 250 μm thick PC film (LEXANTM8010, TEKRA, New Berlin, WI, USA) using a nanostructured nickel master stamp (HT-AR-02A, Temicon, Dortmund, Germany), which features a hexagonally arranged nanostructured pattern with a period of 250 nm and a height of 300 nm. The PC film was imprinted using a hot-pressing machine (QM900M, QMESYS, Uiwang-si, Republic of Korea) by pressing the film between two master molds at elevated temperatures of up to 190 °C for 5 min, with a pressure of 5 MPa for a 5 × 5 cm PC film. To ensure proper formation, cooling was then performed at 100 °C for 10 min before releasing the pressure.

Subsequently, Al_2O_3 was coated on the top side of the thin film using atomic layer deposition (ALD, Atomic-Classic, CN1) at 60 °C. Trimethylaluminum (TMA) and H_2O were used as the precursor and reactant, respectively. The TMA pulse, N_2 purge, H_2O pulse, and N_2 purge cycle were repeated with a base pressure of 500 mTorr and a deposition rate of approximately 0.4 nm/cycle. Finally, SiO_2 was then deposited on the bottom side of the thin film using plasma-enhanced chemical vapor deposition (PECVD, Plasmalab 80+, OXFORD) at 150 °C for 30 s under the following conditions: RF power of 20 W, N_2O flow of 800 sccm, and SiH_4 of 100 sccm.

Optical performance was simulated using rigorous coupled-wave analysis (RCWA) methods to predict reflectance across various refractive indices (air, water, and adhesive) with commercial software (DiffractMod, Rsof t 2021, Synopsys, San Diego, CA, USA). The optical simulations were conducted with hexagonally arranged nanostructures, with a period of 250 nm and height in a visible wavelength. The simulations accounted for varying nanostructure heights, SiO_2, and Al_2O_3 thicknesses for the coating layers to optimize the film's antireflective properties. The optical performances of antireflective films were evaluated by the UV/visible and NIR spectrophotometer (V-770, JASCO, Easton, MD, USA) to measure total reflectance.

Mechanical robustness was evaluated through water contact angle (WCA) measurements using a contact angle analyzer (Phoenix 300, SEO, Suwon-si, Republic of Korea) before and after wear resistance tests. The wear resistance tests included rubbing cycles performed with a rubbing test machine (CT-RB1, Coretech, Incheon, Republic of Korea) equipped with a rubber stick with a durometer A-type hardness of 88 under a 1 kg load. The test films were mounted on a slide glass and positioned on a plate with a path length of 5 cm. After the wear resistance tests, both quantitative and qualitative evaluations were conducted to assess damage to the samples. Scanning electron microscopy (SEM) images were obtained using a Hitachi S-4700 SEM (Tokyo, Japan), while an X-ray photoelectron spectroscopy (XPS, NEXSA, Thermo Fisher Scientific, Waltham, MA, USA) was used to analyze surface morphology and chemical composition changes following mechanical testing.

Bending tests were performed by subjecting the films to different radii of curvature to assess crack formation and durability using a custom setup. The test films were secured on both sides using holders to evaluate the compressive stress applied to the lower side. The compressive stress on the test films was evaluated at radii of curvature of 5.5, 3.5, and 2.5 mm, respectively. Subsequently, surface damage for the bending region was examined using SEM.

The refractive index profile was calculated using numerical software (MATLAB 2016a, MathWorks, Natick, MA, USA). The average refractive index was determined by defining a unit area and calculating the area ratio based on the structure height and coating thickness. Each calculation was performed in 1 nm height increments, and a fixed refractive index value was used to simplify the calculations.

3. Results

3.1. Design and Fabrication of Dual-Coated Antireflective Film

The dual-coated antireflective film was designed by integrating nanostructures on both sides of a thin PC substrate. The film was engineered to address the limitations of traditional antireflective coatings, particularly in terms of flexibility and mechanical robustness. As illustrated in Figure 1a, the top side of the film was coated with a thin layer of Al_2O_3 using ALD. This coating was selected for its high hardness and ability to enhance surface durability, thereby protecting the underlying nanostructures from physical damage during use [22]. The bottom side of the film was coated with SiO_2 via PECVD. The SiO_2 layer was optimized to provide effective antireflective properties, making the film adaptable to various environments, including air, water, and adhesive layers.

Figure 1b presents a schematic illustration comparing the wear resistance of nanostructured PC films with and without an Al_2O_3 coating. SEM images show the surface morphology of the films after 5 rubbing cycles, highlighting the impact of the Al_2O_3 coating on wear resistance. The uncoated nanostructures exhibit significant structural degradation. By contrast, the SEM images of the Al_2O_3-coated film reveal that the nanostructures remain intact, indicating that the Al_2O_3 coating effectively preserves the structural stability of the film during the rubbing test. This evaluation is crucial in ensuring that the mechanical durability of the film is enhanced without sacrificing its optical properties.

To optimize the antireflective performance of the film, we conducted optical simulations to determine the ideal thickness of the SiO_2 coating in relation to the refractive indices of the adjacent layers. Figure 1c presents a reflectance contour plot illustrating the relationship between SiO_2 thickness and the refractive index. The plot demonstrates that a specific SiO_2 thickness can minimize light reflectance across various dynamic environments (e.g., air, water, and adhesive layers). By fine-tuning the SiO_2 thickness, the ARS PC film achieves broad-spectrum antireflective properties that are effective in multiple environments.

Additionally, the refractive index profiles, as shown in Figure 1d, were analyzed as a function of a nanostructure height of 300 nm and SiO_2 coating thickness of 0, 25, and 50 nm. This analysis provided insights into how these parameters influence the antireflective performance of the designed film. The optimized refractive index profiles confirmed that the dual-coated design offers superior optical performance, reducing reflectance to minimal levels across different refractive indices.

The mechanical robustness of the *uni*-ARS PC film was further validated through bending tests. Figure 1e illustrates the schematic comparison of the flexibility between the *uni*-ARS PC film and the flat film. SEM images show that the nanostructured PC film coated with Al_2O_3 and flat PC films with the same coating were both subjected to the bending test. The flat PC film exhibited significant cracks under mechanical stress, highlighting its limited flexibility. In contrast, the nanostructured PC film remained undamaged, demonstrating that the nanostructured design, combined with the Al_2O_3 coating, significantly enhanced the film's mechanical durability and flexibility.

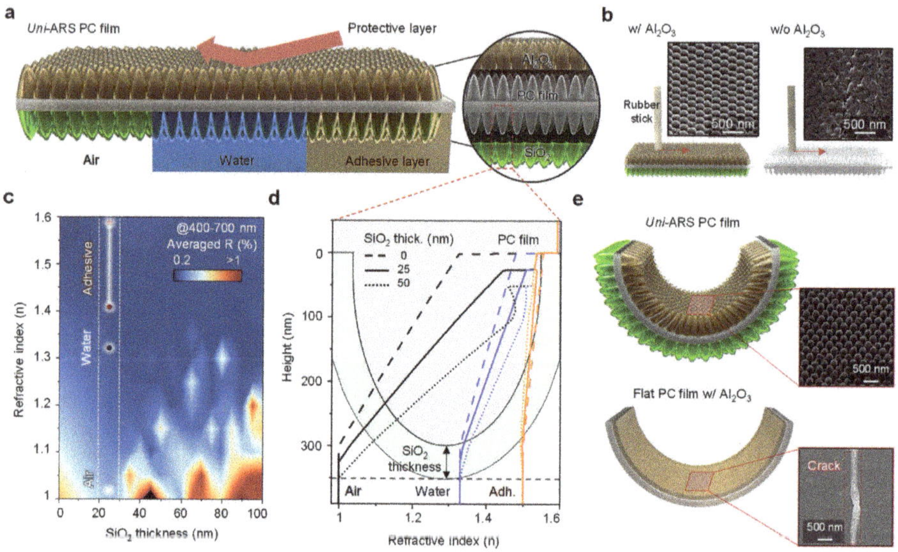

Figure 1. Universal design for antireflective PC film. (**a**) Schematic illustration of dual-coated antireflective film integrating nanostructures on both sides of a PC film. The top side is coated with Al_2O_3 to enhance surface hardness, while the bottom side is coated with SiO_2 to optimize antireflective properties in various environments, including air, water, and adhesive layers. (**b**) A schematic illustration comparing the wear resistance with and without the Al_2O_3 coating. SEM images show the surface morphology, demonstrating the improved durability provided by the coating. (**c**) A reflectance contour plot showing the optical optimization between the SiO_2 thickness and refractive index, highlighting the optimized SiO_2 thickness for various dynamic environments, such as air, water, and adhesive layers. (**d**) Refractive index profiles, as a function of nanostructure height and SiO_2 thickness ((**a**), dash box), illustrating how these parameters influence the antireflective performance of the film. (**e**) A schematic illustration comparing the flexibility of the film with *uni*-ARS and flat PC films. SEM images from bending tests of the *uni*-ARS PC film (top) and flat PC films with Al_2O_3 coating (bottom), indicating that the flat PC film exhibits cracks after the bending test, while the *uni*-ARS PC film remains crack-free.

3.2. Optical Performance and Simulation of the Universal Antireflective Film

The optical performance of the dual-coated antireflective film was thoroughly investigated through both simulations and experimental measurements to assess its effectiveness across various environments with different refractive indices. Figure 2a illustrates the schematic design used for optical simulations, where the PC film is integrated with nanostructures and coated with SiO_2 and Al_2O_3. The period of the nanostructures was set to 250 nm to ensure sufficient transmittance in the visible wavelength range [3]. The SiO_2 layer was specifically tailored to enhance the film's compatibility with environments such as air, water, and adhesive layer, which had a refractive index of 1.5. The thickness of the Al_2O_3 coating was optimized with reflectance and durability. To evaluate the antireflective capabilities of the film, we conducted optical simulations that show averaged reflectance contour plots for visible wavelengths of 400–700 nm, as shown in Figure 2b. The results clearly demonstrate that the film maintains a reflectance below 1% across all tested environments when the ARS height is set at 150 nm. This low reflectance is a critical feature that ensures minimal optical losses, making the film suitable for various optical applications where maintaining high transparency and minimal reflection is essential.

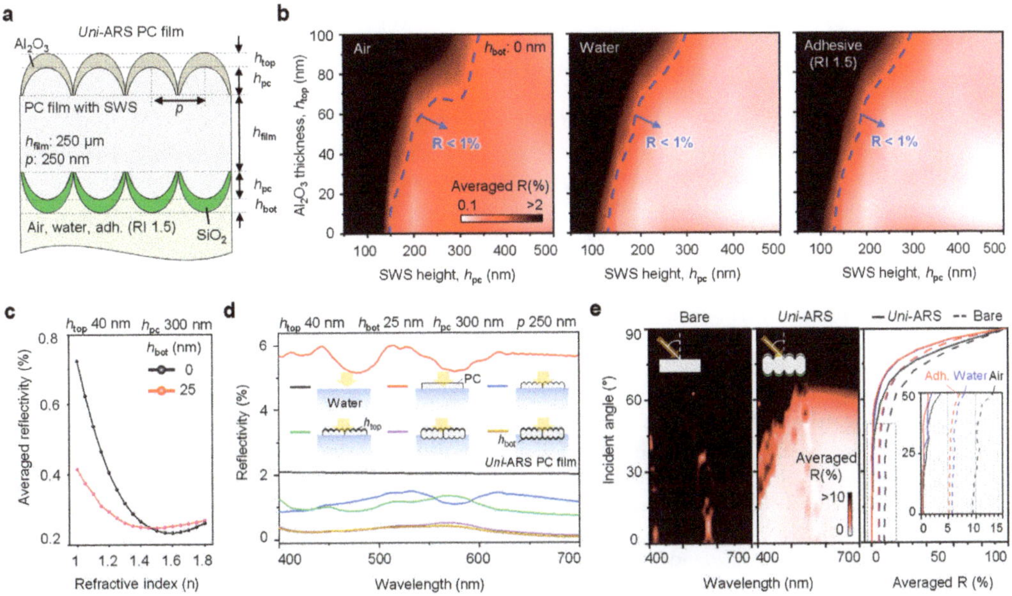

Figure 2. Optical simulation for the universal antireflective PC film. (**a**) A schematic of the PC film with a nanostructure for optical simulation, where the film is coated with SiO_2 for compatibility with air, water, and adhesive environments (refractive index of 1.5). (**b**) Averaged reflectance contour plots for the film in different environments (air, water, and adhesive), showing that the reflectance is maintained below 1% across all cases with an SWS height (h_{pc}) of 150 nm. (**c**) Detailed averaged reflectance profiles as a function of the bottom SiO_2 thickness (h_{bot}) for various environments, demonstrating consistent low reflectance across different refractive indices. (**d**) Schematic and reflectance profiles illustrating how the structure optimizes the optical performance of the proposed film. (**e**) Reflectance comparison as a function of the incident angle between flat and nanostructured PC films, where the flat PC film shows significantly higher reflectance compared to the nanostructured film. The right reflectance plot, according to the incident angle, confirms the consistent superior antireflective performance of the nanostructured film in air, water, and adhesive environments.

Further analysis was conducted to understand how the thickness of the SiO_2 layer (h_{bot}) influences reflectance profiles across different environments. Figure 2c presents the detailed averaged reflectance profiles for visible wavelengths as a function of the bottom SiO_2 thickness. The results indicate that consistent low reflectance can be achieved across a range of refractive indices by optimizing SiO_2 thickness. This consistency underscores the designed film's adaptability and effectiveness in diverse applications, from air-based systems to underwater and adhesive layer environments.

To further optimize the optical performance, we examined how the structure of the nanostructured film impacts reflectance, as depicted in Figure 2d. The schematic and corresponding reflectance profiles illustrate the significance of fine-tuning the nanostructure height and SiO_2 thickness to achieve the best possible antireflective performance. The optimized film structure demonstrates superior optical properties, ensuring that the designed film effectively reduces reflectance across multiple environments without compromising transparency.

Finally, Figure 2e compares the reflectance of flat and *uni*-ARS PC films as a function of the incident angle. The results clearly show that the flat PC film exhibits significantly higher reflectance, particularly at higher incident angles, which is undesirable in many optical applications. In contrast, the nanostructured PC film maintains a low reflectance even as the incident angle increases, confirming its superior antireflective performance. The reflectance plot on the right side further confirms the film's consistent performance in both air and water environments, as well as when applied over adhesive layers, making it highly versatile for a wide range of optical devices.

3.3. Fabrication and Optical Evaluation of the Dual-Coated Antireflective Film

The fabrication process of the dual-coated antireflective film was carefully designed to ensure both the mechanical robustness and superior optical performance of the final film. Figure 3a shows a schematic overview of the fabrication process. The PC film was first imprinted with nanostructures on both sides using a hot-pressing technique, which involved pressing the film between two pattern molds, with a period of 250 nm and a height of 300 nm, at elevated temperatures of up to 190 °C for 5 min with a pressure of 5 MPa [31–33]. After imprinting, one side of the film was coated with Al_2O_3 at a thickness of 40 nm using ALD to enhance surface hardness and durability. The opposite side was coated with SiO_2 with a thickness of 25 nm via PECVD, which was optimized for antireflective performance across various environments. The surface morphology of the nanostructured PC films was examined using SEM images before and after the coating processes. Those of the double-sided ARS PC film displayed the nanostructured surfaces before any coatings were applied, revealing well-defined and uniform nanostructures that are crucial for achieving low reflectance. Those of the Al_2O_3 and SiO_2 coatings showed the surfaces after coating, with Al_2O_3 on the front side and SiO_2 on the back side. The images confirm that the Al_2O_3 and SiO_2 coatings preserved the well-defined shape of the pattern mold after the coating processes, ensuring that the optical and mechanical properties of the film were not degraded.

Figure 3b presents the transmittance spectra measured in the air environment for the fabricated *Uni*-ARS PC film, which corresponds to the fabrication process outlined in Figure 3a. This clearly shows that the *uni*-ARS PC film exhibits enhanced transmittance compared to the film without a SiO_2 coating. This improved transmittance is attributed to a dual-coated nanostructured design, which minimizes reflection and allows for improved light transmission across a visible wavelength. This optical improvement is crucial for applications requiring minimal optical loss, making the *uni*-ARS PC film highly suitable for applications in environments where high optical clarity and low reflectance are essential, such as in displays, solar panels, and transparent electronic devices.

For qualitatively qualifying multi-environmental applications, the optical performance of the dual-coated film was further evaluated under different environmental conditions. Figure 3c presents photographs of the film tested in both air and water environments. When exposed to a white light source, the dual-coated film effectively revealed the underlying logo through the film, both in the air and when submerged in water. This demonstrates the designed film's ability to maintain high transparency and low reflectance across different refractive indices, making it suitable for applications in dynamic environments.

Additionally, Figure 3d highlights the performance of the dual-coated film when applied to a display panel. The film was placed over the display, and its performance was assessed with the display both on and off. The photographs show that the film minimizes reflection and glare, preserving the visibility of the display content under various lighting conditions. When the display is on, the underlying text and images remain clear and distinct, demonstrating that the film effectively reduces light reflection and enhances visibility. The combination of these fabrication techniques and optical evaluations confirms that the dual-coated antireflective film offers significant advantages for both durability and optical performance. The film's ability to adapt to different environments, along with its effectiveness at reducing reflectance, makes it a versatile solution for advanced electronic and optical applications.

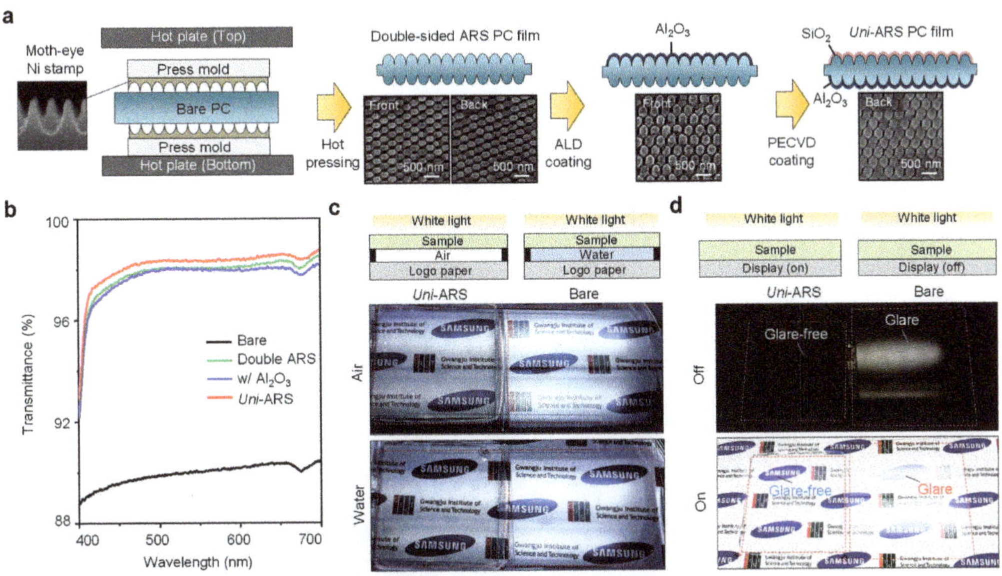

Figure 3. Fabrication and optical evaluation for universal film. (**a**) A schematic illustration of the fabrication process for dual-coated nanostructured PC films. SEM images showing the surface morphology of the nanostructured PC films on the front and back sides, corresponding to each fabrication process. (**b**) Transmittance spectra measured in the air environment for the fabricated film, corresponding to the process shown in Figure 3a, demonstrate that the *uni*-ARS PC film exhibits enhanced transmittance. (**c**,**d**) (**c**) Photographs demonstrating the performance of the dual-coated film under air and water conditions. The *uni*-ARS film under a white light source effectively exhibiting the underlying logo through the film in air and water. (**d**) Photographs demonstrating the performance of the dual-coated film on the display panel. The *uni*-ARS film placed over a display, showing the difference between the display being on (bottom) and off (top). The *uni*-ARS film preserves visibility by suppressing light reflection, effectively reducing glare in both display on and off conditions.

3.4. Mechanical Performance of the Dual-Coated Antireflective Film

To ensure that the dual-coated antireflective film not only excelled in optical performance but also maintained mechanical integrity under various stress conditions, a series of mechanical tests were conducted. These tests evaluated the film's durability, particularly its resistance to mechanical wear and deformation, which are critical factors for practical applications in flexible electronic devices.

Figure 4a presents the results of the WCA measurements, which were used to assess the hydrophobicity and surface integrity of the nanostructured PC films after repeated mechanical stress. The top row shows the WCA measurements for nanostructured films without the Al_2O_3 coating after 30 rubbing cycles. The reduction in WCA indicates a loss of hydrophobicity and surface degradation due to mechanical stress. In contrast, the bottom row shows WCA measurements for nanostructured films with the Al_2O_3 coating. Even after 30 rubbing cycles, the coated film retained a high WCA, indicating that the Al_2O_3 coating effectively preserved the film's hydrophobic properties and surface consistency under mechanical stress.

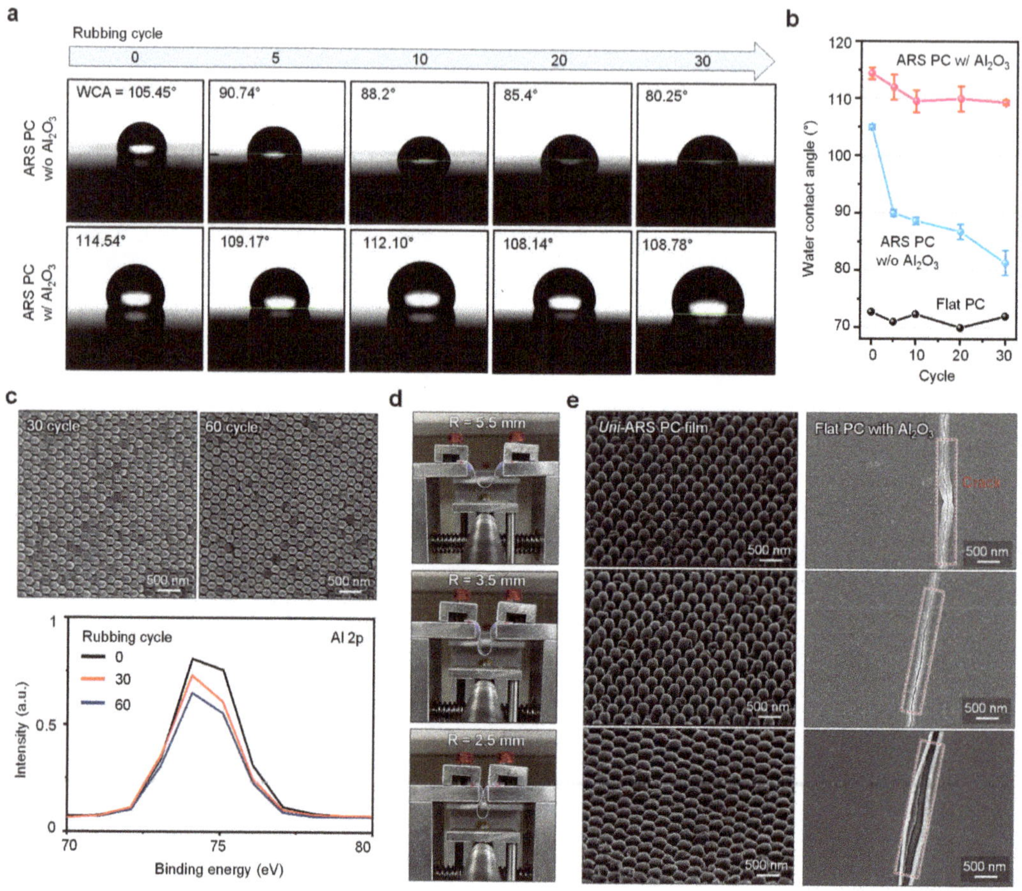

Figure 4. Mechanical performance of the *Uni*-ARS film. (**a**) A comparison of water contact angle measurements for the nanostructured PC films over 30 rubbing tests without the Al_2O_3 coating (top row) and with the Al_2O_3 coating (bottom row). (**b**) Graphs comparing the change in WCA for nanostructured PC films as a function of rubbing cycles in Figure 4a, indicating that the nanostructured film with the Al_2O_3 coating maintains its hydrophobic properties better than the nanostructured PC film without Al_2O_3. To evaluate the stability and reliability of the results, error bars are shown for three samples. (**c**) SEM images of the *uni*-ARS PC film after 30 and 60 rubbing cycles. The bottom plot exhibits XPS data about Al2p according to the rubbing cycle, showing durability. (**d**) Photographs of the bending test setup, showing the *uni*-ARS films under different radii of curvature (R = 5.5 mm, 3.5 mm, 2.5 mm) during the mechanical test. (**e**) SEM images comparing the *uni*-ARS PC film and flat PC film with the Al_2O_3 coating after the bending test, revealing cracks on the flat PC film while the *uni*-ARS PC film remained intact.

Figure 4b provides a graphical comparison of the change in WCA for both coated and uncoated nanostructured PC films as a function of the number of rubbing cycles shown in Figure 4a. The graph shows that while the WCA of the uncoated film decreases significantly with increasing rubbing cycles, the coated film exhibits only a minimal decrease, further confirming the protective role of the Al_2O_3 coating in maintaining the film's hydrophobicity and mechanical durability. Furthermore, to evaluate the stability and reliability of the mechanical properties of the Al_2O_3 coatings, we measured the WCA for three samples with and without Al_2O_3 coatings, respectively. In all samples, the Al_2O_3-coated film consistently showed a higher contact angle compared to the uncoated film for overall rubbing cycles.

The durability of the film's nanostructures under mechanical stress was further examined using SEM imaging after 30 and 60 rubbing cycles, as shown in Figure 4c. The SEM images demonstrate that the nanostructures remain largely undamaged after 30 cycles, with only minor wear observed after 60 cycles. Additionally, XPS analysis of the Al2p peak was conducted, with the results presented in the bottom plot of Figure 4c. The XPS data indicate that the Al_2O_3 coating remained stable and effectively bonded to the nanostructured surface even after multiple rubbing cycles, highlighting the coating's durability.

To assess the film's flexibility and resistance to cracking under deformation, bending tests were conducted, as illustrated in Figure 4d. The photographs show the film being stressed to different radii of curvature (R = 5.5 mm, 3.5 mm, and 2.5 mm) during the bending tests. The results reveal that the dual-coated film can withstand significant bending without visible damage or cracking, demonstrating its suitability for flexible electronic applications.

Finally, Figure 4e compares SEM images of the dual-coated nanostructured PC film with Al_2O_3 to those of a flat PC film with the same coating after the bending test. The flat PC film exhibits significant cracking, compromising its mechanical stability. In contrast, the designed film remains free of cracks, highlighting the superior mechanical robustness provided by the designed nanostructure and protective Al_2O_3 coating. This result underscores the effectiveness of the dual-coating strategy in enhancing both the optical and mechanical properties of the film, making it an ideal candidate for flexible, durable antireflective applications.

4. Conclusions

Moth-eye-inspired artificial antireflective nanostructured surfaces are highly effective at reducing optical losses at interfaces, offering significant benefits for enhancing optical performance in various optoelectronic applications. Despite these advantages, their broader use is constrained by limited flexibility and low surface hardness. In this study, we successfully developed a dual-coated antireflective film with multifunctional applications of flexibility and robust mechanical and optical performance, making it suitable for multi-environmental optoelectronic applications. By integrating nanostructures on both sides of a thin PC film and applying Al_2O_3 and SiO_2 coatings, we achieved a *uni*-ARS PC film that consistently reduced reflectance across various refractive indices while maintaining durability under mechanical stress (e.g., compressive strain, wear resistance). The experimental results and simulations demonstrate that this advanced approach effectively addresses the limitations of traditional antireflective surfaces, offering a versatile and durable solution for advanced optical technologies.

Our *uni*-ARS PC film demonstrated remarkable performance, maintaining low reflectance across a range of refractive indices while also exhibiting superior mechanical performance, particularly with better wear resistance in flexible electronic applications. The robust and flexible nature of our film makes it an ideal candidate for use in flexible displays, solar panels, and other optoelectronic devices where both durability and optical clarity are paramount. Future research could explore a broader range of conditions and configurations to further validate the universality of the film's performance. In addition, future studies could investigate alternative materials for the coating layers to further enhance specific properties, such as hydrophobicity or thermal stability. Furthermore, scaling up the fabrication process for industrial applications and testing the film's performance under more extreme environmental conditions would be valuable for advanced technologies.

Author Contributions: Conceptualization, H.J.J. and Y.M.S.; Data curation, H.J.J. and J.H.Y.; Formal analysis, H.J.J., J.J., J.H.Y., H.L. and S.H.; Methodology, H.J.J. and Y.M.S.; Resources, H.J.J., J.J., S.K. and D.K.; Software, H.J.J.; Supervision, Y.M.S.; Validation, H.J.J., I.S.T. and S.A.; Visualization, H.J.J.; Writing—original draft, H.J.J.; Writing—review and editing, H.J.J., J.J., I.S.T. and Y.M.S. All authors have read and agreed to the published version of the manuscript.

Funding: This work was supported by Samsung Display Co., Ltd. This work was supported also by the National Research Foundation of Korea (NRF) grant funded by the Korean government (MSIT) (NRF-2022M3H4A1A02046445, NRF-2021M3H4A1A04086357).

Institutional Review Board Statement: Not applicable.

Data Availability Statement: The data that support the plots within this paper and other findings of this study are available on request from the corresponding author upon reasonable request. The data are not publicly available due to funders' policies.

Conflicts of Interest: The authors declare no conflicts of interest.

References

1. Yeo, C.; Kim, J.B.; Song, Y.M.; Lee, Y.T. Antireflective silicon nanostructures with hydrophobicity by metal-assisted chemical etching for solar cell applications. *Nanoscale Res. Lett.* **2013**, *8*, 1–7. [CrossRef] [PubMed]
2. Rahman, A.; Ashraf, A.; Xin, H.; Tong, X.; Sutter, P.; Eisaman, M.D.; Black, C.T. Sub-50-nm self-assembled nanotextures for enhanced broadband antireflection in silicon solar cells. *Nat. Commun.* **2015**, *6*, 5963. [CrossRef] [PubMed]
3. Jang, H.J.; Kim, Y.J.; Yoo, Y.J.; Lee, G.J.; Kim, M.S.; Chang, K.S.; Song, Y.M. Double-sided anti-reflection nanostructures on optical convex lenses for imaging applications. *Coatings* **2019**, *9*, 404. [CrossRef]
4. Cai, J.; Qi, L. Recent advances in antireflective surfaces based on nanostructure arrays. *Mater. Horiz.* **2015**, *2*, 37–53. [CrossRef]
5. Kang, E.K.; Kwon, E.; Min, J.W.; Song, Y.M.; Lee, Y.T. Improved light extraction efficiency of GaN-based vertical LEDs using hierarchical micro/subwavelength structures. *Jpn. J. Appl. Phys.* **2015**, *54*, 06FH02. [CrossRef]
6. Ko, J.H.; Kim, S.H.; Kim, M.S.; Heo, S.Y.; Yoo, Y.J.; Kim, Y.J.; Lee, H.; Song, Y.M. Lithography-free, large-area spatially segmented disordered structure for light harvesting in photovoltaic modules. *ACS Appl. Mater. Interfaces* **2022**, *14*, 44419–44428. [CrossRef]
7. Kim, Y.J.; Yoo, Y.J.; Yoo, D.E.; Lee, D.W.; Kim, M.S.; Jang, H.J.; Kim, Y.C.; Jang, J.H.; Kang, I.S.; Song, Y.M. Enhanced light harvesting in photovoltaic devices using an edge-located one-dimensional grating polydimethylsiloxane membrane. *ACS Appl. Mater. Interfaces* **2019**, *11*, 36020–36026. [CrossRef]
8. Liu, S.; Tso, C.Y.; Lee, H.H.; Zhang, Y.; Yu, K.M.; Chao, C.Y.H. Bio-inspired TiO_2 nano-cone antireflection layer for the optical performance improvement of VO_2 thermochromic smart windows. *Sci. Rep.* **2020**, *10*, 11376. [CrossRef]
9. Choi, C.; Hinton, H.; Seung, H.; Chang, S.; Kim, J.S.; You, W.; Kim, M.S.; Hong, J.P.; Lim, J.A.; Hwang, D.K.; et al. Anti-distortion bioinspired camera with an inhomogeneous photo-pixel array. *Nat. Commun.* **2024**, *15*, 6021. [CrossRef]
10. Kim, D.H.; Kim, J.J.; Kong, D.-J.; Lee, G.J.; Song, Y.M. Bio-inspired tunable optics and photonics: Bridging the gap between nature and technology. *Int. J. Optomechatron.* **2024**, *18*, 2334293. [CrossRef]
11. Chang, S.; Koo, J.H.; Yoo, J.; Kim, M.S.; Choi, M.K.; Kim, D.-H.; Song, Y.M. Flexible and Stretchable Light-Emitting Diodes and Photodetectors for Human-Centric Optoelectronics. *Chem. Rev.* **2024**, *124*, 768–859. [CrossRef] [PubMed]
12. Sikder, U.; Zaman, M.A. Optimization of multilayer antireflection coating for photovoltaic applications. *Opt. Laser Technol.* **2016**, *79*, 88–94. [CrossRef]
13. Kumaragurubaran, B.; Anandhi, S. Reduction of reflection losses in solar cell using anti reflective coating. In Proceedings of the 2014 International Conference on Computation of Power, Energy, Information and Communication (ICCPEIC), Chennai, India, 16–17 April 2014; IEEE: Piscataway, NJ, USA, 2014; pp. 155–157.
14. Paulson, C.A.; Price, J.J.; Koch, K.W.; Kim, C.G.; Oh, J.H.; Lin, L.; Subramanian, A.N.; Zhang, B.; Amin, J.; Mayolet, A.; et al. Industrial-grade anti-reflection coatings with extreme scratch resistance. *Opt. Lett.* **2019**, *44*, 5977–5980. [CrossRef] [PubMed]
15. Song, Y.M.; Jang, S.J.; Yu, J.S.; Lee, Y.T. Bioinspired parabola subwavelength structures for improved broadband antireflection. *Small* **2010**, *6*, 984–987. [CrossRef]
16. Liu, K.; Yao, X.; Jiang, L. Recent developments in bio-inspired special wettability. *Chem. Soc. Rev.* **2010**, *39*, 3240–3255. [CrossRef]
17. Qu, C.; Rozsa, J.L.; Jung, H.J.; Williams, A.R.; Markin, E.K.; Running, M.P.; McNamara, S.; Walsh, K.M. Bio-inspired antimicrobial surfaces fabricated by glancing angle deposition. *Sci. Rep.* **2023**, *13*, 207. [CrossRef]
18. Kelleher, S.M.; Habimana, O.; Lawler, J.; O'Reilly, B.; Daniels, S.; Casey, E.; Cowley, A. Cicada Wing Surface Topography: An Investigation into the Bactericidal Properties of Nanostructural Features. *ACS Appl. Mater. Interfaces* **2016**, *8*, 14966–14974. [CrossRef]
19. Nam, W.I.; Yoo, Y.J.; Song, Y.M. Geometrical shape design of nanophotonic surfaces for thin film solar cells. *Opt. Express* **2016**, *24*, A1033–A1044. [CrossRef]
20. Choi, K.; Yoon, Y.; Jung, J.; Ahn, C.W.; Lee, G.J.; Song, Y.M.; Ko, M.J.; Lee, H.S.; Kang, I.S. Super-Antireflective Structure Films with Precisely Controlled Refractive Index Profile. *Adv. Opt. Mater.* **2017**, *5*, 3. [CrossRef]
21. Jacobo-Martín, A.; Hernández, J.J.; Solano, E.; Monclús, M.A.; Martínez, J.C.; Fernandes, D.F.; Pedraz, P.; Molina-Aldareguia, J.M.; Kubart, T.; Rodríguez, I. Resilient moth-eye nanoimprinted antireflective and self-cleaning TiO_2 sputter-coated PMMA films. *Appl. Surf. Sci.* **2022**, *585*, 152653. [CrossRef]
22. Yoo, Y.J.; Kim, Y.J.; Kim, S.Y.; Lee, J.H.; Kim, K.; Ko, J.H.; Lee, J.W.; Lee, B.H.; Song, Y.M. Mechanically robust antireflective moth-eye structures with a tailored coating of dielectric materials. *Opt. Mater. Express* **2019**, *9*, 4178–4186. [CrossRef]

23. Isakov, K.; Kauppinen, C.; Franssila, S.; Lipsanen, H. Superhydrophobic antireflection coating on glass using grass-like alumina and fluoropolymer. *ACS Appl. Mater. Interfaces* **2020**, *12*, 49957–49962. [CrossRef] [PubMed]
24. Sebastian, D.; Yao, C.W.; Lian, I. Abrasion resistance of superhydrophobic coatings on aluminum using PDMS/SiO_2. *Coatings* **2018**, *8*, 414. [CrossRef]
25. Chiou, A.H.; Chang, C.W.; Ting, C.J. Spectrally selective antireflection of nanoimprint lithography-formed 3D spherical structures on film coated with a silver layer. *Sci. Rep.* **2022**, *12*, 19505. [CrossRef]
26. Kim, M.; Lee, G.J.; Choi, C.; Kim, M.S.; Lee, M.; Liu, S.; Cho, K.W.; Kim, H.M.; Cho, H.; Choi, M.K.; et al. An Aquatic Vision Inspired Camera based on a Monocentric Lens and a Silicon Nanorod Photodiode Array. *Nat. Electron.* **2020**, *3*, 546–553. [CrossRef]
27. Song, Y.M.; Xie, Y.; Malyarchuk, V.; Xiao, J.; Jung, I.; Choi, K.J.; Liu, Z.; Park, H.; Lu, C.; Kim, R.H.; et al. Digital cameras with designs inspired by the arthropod eye. *Nature* **2013**, *497*, 95–99. [CrossRef]
28. Kim, M.; Chang, S.; Kim, M.; Yeo, J.; Kim, M.S.; Lee, G.J.; Kim, D.H.; Song, Y.M. Cuttlefish-eye-inspired artificial vision for high-quality imaging under uneven illumination conditions. *Sci. Robot.* **2023**, *8*, eade4698. [CrossRef]
29. Lee, M.; Lee, G.J.; Jang, H.J.; Joh, E.; Cho, H.; Kim, M.S.; Kim, H.M.; Kang, K.M.; Lee, J.H.; Kim, M.; et al. An amphibious artificial vision system with a panoramic visual field. *Nat. Electron.* **2022**, *5*, 452–459. [CrossRef]
30. Park, J.; Kim, M.S.; Kim, J.; Chang, S.; Lee, M.; Lee, G.J.; Song, Y.M.; Kim, D.H. Avian eye–inspired perovskite artificial vision system for foveated and multispectral imaging. *Sci. Robot.* **2024**, *9*, eadk6903. [CrossRef]
31. Yeo, N.E.; Cho, W.K.; Kim, D.I.; Jeong, M.Y. Enhanced anti-scratch performance of nanopatterned anti-reflective polymer films. *Appl. Surf. Sci.* **2018**, *458*, 503–511. [CrossRef]
32. Park, S.; Kim, J.; Han, T.W.; Hwang, D.Y.; Lee, H.C.; Kim, W.B. Mechanical reinforcement of UV-curable polymer nanocomposite for nanopatterned mold. *Microelectron. Eng.* **2022**, *259*, 111791. [CrossRef]
33. Xiao, Y.; Cao, X.; Dong, Q.; Gao, L. All dissolvable and transient plasmonic device enabled by nanoimprint lithography. *Nanotechnology* **2023**, *34*, 295301. [CrossRef]

biomimetics

Article

Eye-Inspired Single-Pixel Imaging with Lateral Inhibition and Variable Resolution for Special Unmanned Vehicle Applications in Tunnel Inspection

Bin Han [1], Quanchao Zhao [2], Moudan Shi [1,2], Kexin Wang [3], Yunan Shen [1], Jie Cao [1,2,*] and Qun Hao [1,2,4]

1 School of Optics and Photonics, Beijing Institute of Technology, Beijing 100081, China; 3120170330@bit.edu.cn (B.H.); 3220220495@bit.edu.cn (M.S.); 3120240572@bit.edu.cn (Y.S.); qhao@bit.edu.cn (Q.H.)
2 Yangtze Delta Region Academy of Beijing Institute of Technology, Jiaxing 314019, China; zqc@bitjx.edu.cn
3 The China Railway 12th Bureau Group Company, Ltd., Taiyuan 030027, China; tiehang@yftong.com.cn
4 Changchun University of Science and Technology, Changchun 130022, China
* Correspondence: caojie@bit.edu.cn

Abstract: This study presents a cutting-edge imaging technique for special unmanned vehicles (UAVs) designed to enhance tunnel inspection capabilities. This technique integrates ghost imaging inspired by the human visual system with lateral inhibition and variable resolution to improve environmental perception in challenging conditions, such as poor lighting and dust. By emulating the high-resolution foveal vision of the human eye, this method significantly enhances the efficiency and quality of image reconstruction for fine targets within the region of interest (ROI). This method utilizes non-uniform speckle patterns coupled with lateral inhibition to augment optical nonlinearity, leading to superior image quality and contrast. Lateral inhibition effectively suppresses background noise, thereby improving the imaging efficiency and substantially increasing the signal-to-noise ratio (SNR) in noisy environments. Extensive indoor experiments and field tests in actual tunnel settings validated the performance of this method. Variable-resolution sampling reduced the number of samples required by 50%, enhancing the reconstruction efficiency without compromising image quality. Field tests demonstrated the system's ability to successfully image fine targets, such as cables, under dim and dusty conditions, achieving SNRs from 13.5 dB at 10% sampling to 27.7 dB at full sampling. The results underscore the potential of this technique for enhancing environmental perception in special unmanned vehicles, especially in GPS-denied environments with poor lighting and dust.

Keywords: single-pixel imaging; variable resolution; lateral inhibition; tunnel inspection; unmanned vehicles

1. Introduction

In the realm of public safety and infrastructure maintenance, the deployment of special unmanned vehicles (UAVs) has become increasingly vital, particularly in GPS-denied environments such as tunnel inspection and bridge monitoring [1–3]. These operations, which are essential for public safety, require frequent and accurate assessments to ensure the integrity and safety of critical infrastructure. However, traditional imaging techniques often struggle in complex environments characterized by poor lighting and the presence of dust [1,4]. The constraints on the size and weight of imaging systems for unmanned vehicles further exacerbate these challenges, demanding a novel approach to optical imaging that is both compact and powerful [5].

Single-pixel imaging (SPI) has emerged as an innovative solution to these limitations, offering structural simplicity, resolution independence, and high sensitivity [6–8]. This technology is particularly adept at operating in low-light conditions, presenting a promising new path for enhancing the environmental perception capabilities of UAVs. To address the challenges faced in tunnel inspection, we have partnered with the China Railway 12th Bureau Group Company, Ltd., to develop an eye-inspired single-pixel imaging system with lateral inhibition and variable resolution. This system is designed to significantly enhance a vehicle's environmental perception capabilities in challenging environments.

Our approach integrates variable-resolution imaging with lateral inhibition, aiming to improve the efficiency and quality of the image reconstruction of fine targets in practical applications. This technique mimics the human eye's ability to concentrate high-resolution vision only in the region of interest, offering significant advancement in the field of imaging for unmanned vehicle navigation and inspection tasks. By emulating the high-resolution foveal vision of the human eye, this method significantly enhances the efficiency and quality of image reconstruction for fine targets within the region of interest (ROI). This method utilizes non-uniform speckle patterns coupled with lateral inhibition to augment optical nonlinearity, leading to superior image quality and contrast. Lateral inhibition effectively suppresses background noise, thereby improving the imaging efficiency and substantially increasing the signal-to-noise ratio (SNR) in noisy environments [2,9].

The concept of single-pixel imaging is not new; it has been explored extensively in the field of quantum imaging and has recently transitioned to classical computational imaging [6,10,11]. The unique advantage of GI lies in its ability to separate the detection of light from the formation of an image, allowing the use of a single-pixel detector and enabling high-speed, high-resolution imaging [12–14]. This is particularly beneficial in scenarios where the imaging environment is hostile or where the target is moving rapidly [7,8,15].

In this study, we present a cutting-edge imaging technique that integrates single-pixel imaging inspired by the human visual system with lateral inhibition and variable resolution to improve environmental perception in challenging conditions, such as those with poor lighting and dust. By emulating the high-resolution foveal vision of the human eye, this method significantly enhances the efficiency and quality of image reconstruction for fine targets within the region of interest (ROI). This method utilizes non-uniform speckle patterns coupled with lateral inhibition to augment optical nonlinearity, leading to superior image quality and contrast. Lateral inhibition effectively suppresses background noise, thereby improving the imaging efficiency and substantially increasing the signal-to-noise ratio (SNR) in noisy environments [2,9].

The structure of this paper is as follows: The subsequent section delves into the principle of our eye-inspired GI, detailing the theoretical underpinnings and the innovative aspects of our approach. We then describe the methods employed in our study, including the experimental setup and the protocol for data acquisition and analysis. The results section presents the findings from our indoor experiments and field tests, demonstrating the performance of our method under controlled conditions and in real-world tunnel scenarios. Finally, we conclude with a discussion of the implications of our results, the potential applications of this technology, and the future directions of our research.

Through this work, we aim to contribute a reliable imaging solution that enhances safety and efficiency in tunnel inspections and other applicable domains, marking a significant advancement over traditional imaging techniques. Our research underscores the potential of bio-inspired computational imaging to overcome the limitations of current technologies and to meet the demanding requirements of special unmanned vehicles in GPS-denied environments.

2. Methods

2.1. Principle

The principle of eye-inspired GI [16,17] is shown in Figure 1, in which the laser is triggered by the main board and illuminates the target through a beam expander. The light is then reflected from the target and modulated by the digital mirror device (DMD) [18,19]. The DMD is loaded with our pre-designed variable-resolution speckles. The square where the target object is located is modulated by a variable-resolution speckle. A single-pixel detector receives light for modulation [20,21]. Finally, the target image is reconstructed and displayed by performing a cross-correlation operation [22] between the light intensity signals and the light field information modulated by the DMD.

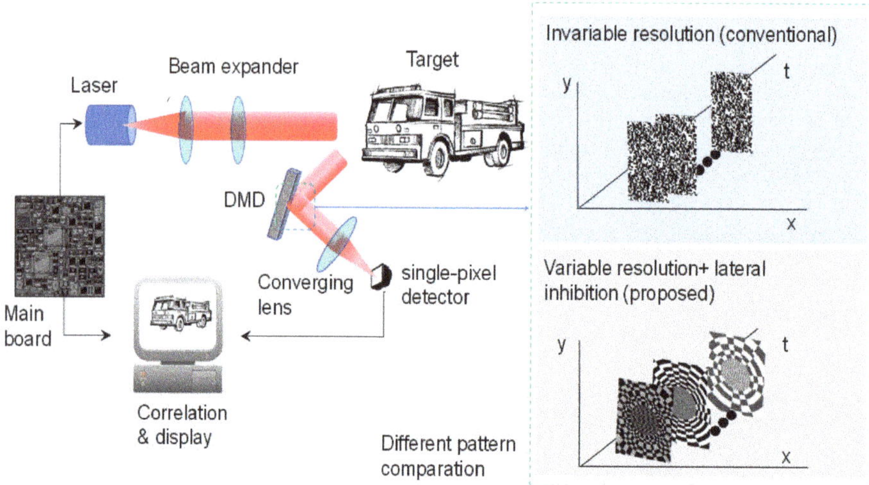

Figure 1. The principle of eye-inspired GI.

From the principle above, the marketable difference of our method is the use of non-uniform patterns combined with the feature of lateral inhibition. Therefore, the advantages of the proposed method include the following: (1) Enhanced Image Quality: The non-uniform speckle pattern increases the optical nonlinearity of the imaging system, which aids in capturing more details and contrast during the imaging process, thereby enhancing image quality. (2) Improved Imaging Efficiency: The lateral inhibition mechanism suppresses background noise and enhances the useful signal, thus improving the imaging efficiency of the system. (3) Improved noise resistance: The lateral inhibition mechanism helps reduce noise interference during the imaging process, particularly in complex or noisy imaging environments; thus, it can significantly enhance the SNR of the imaging.

2.2. Variable-Resolution Projection Pattern

In this study, a new variable-resolution projection pattern is used to replace the traditional fixed-resolution projection pattern. The design of this new pattern is inspired by the non-uniform distribution pattern of receptors on the retina of the human eye, where the projection area is divided into a high-resolution fovea region and a low-resolution marginal region. In the fovea region, we use the standard uniform high-resolution Cartesian sampling method, which avoids the oversampling problem that can occur when using log-polar coordinates in the central region. In the edge region, we use the logarithmic polar coordinate variable-resolution sampling technology, using the characteristics of this coordinate system to compress the image properly. With this design, we successfully simulated the imaging characteristics of the human eye, which provides a high-resolution image in the central area of the field of view and a low-resolution image in the peripheral area of the field of view.

The variable-resolution projection pattern consists of two parts, the foveal region and the edge region, as shown in Figure 2. Figure 2a shows a detailed enlarged view of the edge region; Figure 2b shows the entire variable-resolution projection pattern, including the foveal region and the marginal region; Figure 2c specifically shows the foveal area; Figure 2d shows the marginal region. In the fovea, high-resolution sampling is generated using the traditional Cartesian coordinate system, which avoids the oversampling problem that occurs when using log-polar coordinates in the fovea. The edges are based on a log-polar model that mimics the human eye. If r is used to represent the distance from the pixel point to the center point, then the outer boundary of the foveal region, that is, the ring with radius r_0, belongs to the foveal region; and the part outside this ring is the edge region.

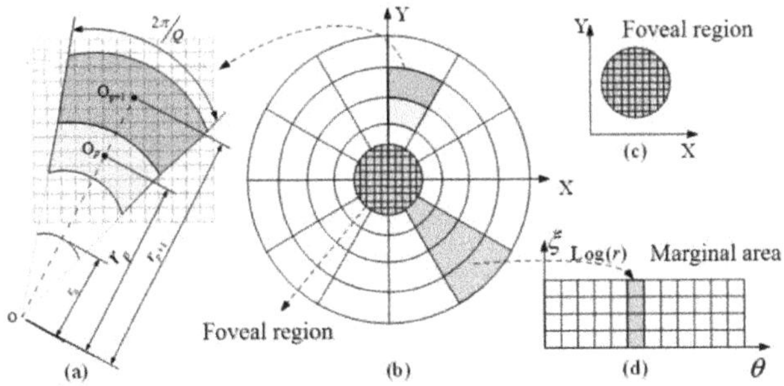

Figure 2. Variable-resolution projection pattern structure diagram.

According to the difference between the polar diameter and the polar angle, the edge region can be divided into P rings; each ring has Q pixels. In the human-simulated log-polar model, let p and q represent the p-ring and QTH pixel, respectively; then, the variable-resolution structure of the edge region can be calculated by the following equations:

$$\begin{cases} r_{p+1} = r_1 \cdot \varepsilon^p \\ \varepsilon = \frac{1+\sin(\pi/Q)}{1-\sin(\pi/Q)} \\ r_1 = \frac{r_0}{1-\sin(\pi/Q)} \\ \theta_q = q \cdot \frac{2\pi}{Q} \qquad (q = 1,2,3\ldots Q) \\ \xi = log_\varepsilon(r_p) = log_\varepsilon(r_1) + p - 1 \qquad (p = 1,2,3\ldots P) \end{cases} \tag{1}$$

where r_p represents the radius of the ring where the PTH ring pixel is located, θ_q represents the angle corresponding to the q pixel, and ε represents the growth coefficient between the rings.

The steps to achieve the variable-resolution projection pattern are as follows: First, set the radius of the foveal area r_0 as a reference. Then, according to the distance between each pixel and the center point, the whole projection pattern is divided into the foveal region and the variable-resolution edge region. In the foveal region, we divide it evenly into P_f pixels according to the conventional method. For the variable-resolution edge region, we calculate the additional parameters required for generation according to specific constraints. For example, if the constraint is the number of pixels per ring Q, then we calculate the growth coefficient between rings ε, the radius of each ring rp, and the number of rings P based on this condition. Then, Formula (1) is used to generate P_p pixels of the edge region. Similarly, other required parameters can be calculated from the number of rings P. Finally, a complete variable-resolution projection pattern is obtained by merging the foveal region with the variable-resolution edge region.

2.3. The SPI with Variable Resolution and Lateral Inhibition

The core theoretical contribution of this method is the integration of variable resolution and lateral inhibition into a single-pixel imaging model. The model mimics the foveated vision of the human eye, where high-resolution imaging is reserved for the region of interest, whereas the periphery is imaged at a lower resolution [23,24]. Meanwhile, lateral inhibition is incorporated to enhance contrast and detail in the final image reconstruction. The image reconstruction based on the SPI pattern model of variable resolution and lateral inhibition includes six steps.

Step 1: Variable resolution modulation function

Variable Resolution Concept: Variable resolution adjusts imaging detail based on the importance of different areas within the field of view (FOV), with high resolution for the region of interest (ROI) and lower resolution for the peripheral areas. The modulation function $M(x, y)$ is expressed as:

$$M(x,y) = exp\left[-\frac{d^2(x,y)}{2\sigma^2}\right] \tag{2}$$

where $d(x, y)$ is the distance from point (x, y) to the center of the FOV, and σ is a decay parameter that controls the sharpness of the transition between high- and low-resolution areas.

Step 2: Modulation of the Speckle Pattern

Based on Step 1, the integration of the variable resolution $P(x, y)$ is written as

$$P(x,y) = P_0(x,y)\cdot M(x,y) \tag{3}$$

where $P_0(x, y)$ is the initial uniform speckle pattern and $M(x, y)$ modulates this pattern based on the variable resolution requirement, thereby enhancing the resolution of the ROI.

Step 3: Application of Lateral Inhibition

We introduce the lateral inhibition function $S(x, y)$, which is written as

$$S(x,y) = \begin{cases} 1 & if\ (x,y)\ is\ in\ the\ ROI \\ \varepsilon & otherwise \end{cases} \tag{4}$$

where ε is a small positive number. This function is crucial for enhancing the contrast by reducing the effect of the peripheral region.

Step 4: Generation of the Detected Beam

The intensity of the detected beam's intensity $D(x, y)$ after passing through the target is given by

$$D(x,y) = \iint P(x',y')\cdot I(x'-x,\ y'-y)dx'dy' \tag{5}$$

where $P(x', y')$ is the modulated speckle pattern that considers both variable resolution and lateral inhibition, and $I(x' - x, y' - y)$ is the intensity distribution of the target.

Step 5: Calculation of the second-order correlation function

The second-order correlation function $G^2(\tau)$ is used in the imaging process and is written as

$$G^2(\tau) = \left\langle I_{ref}(t)I(t+\tau)\right\rangle \tag{6}$$

This function measures the correlation between the reference beam and the detected beam, which is influenced by the modulated speckle pattern.

Step 6: Image Reconstruction

$$I'(x,y) \propto \iint G^2(\tau)\cdot\left[P_0(x'+x,\ y'+y)\cdot M(x',y')\cdot S(x',y')\right] \times dx'dy' \tag{7}$$

The reconstructed image $I'(x, y)$ is proportional to the integral of the product of the second-order correlation function and the complex conjugate of the modulated speckle pattern, which includes both variable-resolution and lateral inhibition effects.

From the above processes, by integrating the variable-resolution model more explicitly into the derivation, we enhance the clarity of how different areas within the FOV are treated based on their importance. This approach ensures that the ROI is imaged with a higher resolution and contrast, while peripheral areas are adjusted to reduce noise and interference, leading to a more effective single-pixel imaging technique.

According to the size of the region of interest and the region of non-interest, the log-polar model can be used to generate the spatial variable-resolution sampling structure with different position resolutions. By using the side suppression feature, the projection pattern can be optimized based on the boundary and detail characteristics of the obtained imaging results. In traditional ghost imaging, random speckles or stripes based on image transformation basis are used as projection patterns. However, the characteristics of the current imaging scene are not considered, especially for the low-contrast scene which is poorly imaged by the existing methods. The method proposed in this paper establishes the connection between the scene information and the projection pattern by combining the side suppression feature with the variable-resolution speckle and adaptively adjusts the projection pattern according to the characteristics of the current scene, which can effectively enhance the edge details and improve the contrast and image quality.

3. Experiments and Results

The experiments were designed to validate two critical aspects of the proposed method: the compression of redundant data and the ability to capture clear images under actual working conditions. To address these objectives, both indoor and field experiments were conducted. The indoor experiments utilized discrete optical components to validate the advantages of the compression of redundant data, whereas the field experiments were conducted in real tunnel scenarios to mimic real-world operational challenges.

3.1. Experimental Setup

In the indoor experiments, as shown in Figure 3, we assembled a high-performance system that corresponded to the components diagrammatically represented in the provided image. The heart of our setup was a pulse laser (①) from the DSS 1064 series of the CryLas laser system, selected for its stability and precision to ensure reliable light emission. This was complemented by a Vialux V7001 DMD (②), chosen for its 1024 $\times$ 768 resolution and 22 kHz micromirror flip rate, to generate and modulate speckle patterns. The DMD was securely mounted adjacent to the pulse laser, with light-shielding tape applied to prevent stray light interference. Our transmission optical component (③) consisted of a series of precision optical elements, such as lenses and beam shapers that direct and focus the speckle pattern onto the detection target (④) [25,26]. The targets, including triangular, circular, and square shapes, were fabricated using 3D printing technology to construct a 3D scene for imaging, simulating potential tunnel-inspection scenarios. The light reflected from the targets was directed to two single-pixel photodiodes (⑥) [16,27] via a beam splitter (⑤). The photodiode (Thorlabs' PDA36E, Sorebo Optoelectronic Technology Shanghai, China), recognized for its high sensitivity and wide bandwidth, effectively captured the nuances of the light signals. All these components were mounted on custom 3D-printed bases to ensure accurate alignment. To manage the overall timing of the system, which is crucial for measuring the time intervals of the reflected light signals and facilitating the reconstruction of single-pixel images, we employed a field-programmable gate array (FPGA)-based time-to-digital converter (TDC) system. A data acquisition card (GaGe CSE22G8) was used to collect and digitize the analog signals from the photodiodes, which were then transferred to a computer for further processing and analysis. This comprehensive setup, with its carefully selected and integrated components, allowed us to effectively evaluate the variable-resolution imaging technique under controlled conditions, demonstrating its potential to enhance image quality, especially in capturing fine details within the ROI.

In the indoor experiments, we utilized a setup that included geometric targets, such as a triangular prism with a 10 cm side length, a circular disc with a 10 cm diameter, and a square prism with a 10 cm side length. These targets were strategically positioned 60 cm from the light source.

In the field experiments, we aimed to address the practical challenges of tunnel inspection by collaborating with the China Railway Twelve Bureau to develop a specialized integrated land and air inspection robot. The robot was designed to navigate and inspect the interior walls of tunnels according to pre-planned paths and to transmit the collected inspection data to a terminal for analysis by professionals. Given that the tunnels under inspection are not in their final state, the robot operates in harsh environments characterized by GPS signal blockage, dim lighting, and complex obstacles on the tunnel walls. These conditions make it difficult for the robot to avoid obstacles; the typical obstacles include wires, cables, and angle irons. In the experiments, we selected a cable of 0.5 cm as the target. While existing 3D imaging LiDAR can map the interior of a tunnel and roughly locate larger obstacles, it struggles to accurately image small, fine details; such imaging is crucial for precise localization. Additionally, the size, weight, and power consumption of current LiDAR systems make it challenging to integrate them directly into special unmanned vehicles. We focused on field experiments that imaged fine targets under actual working conditions to guide and enhance the environmental perception capabilities of special unmanned vehicles.

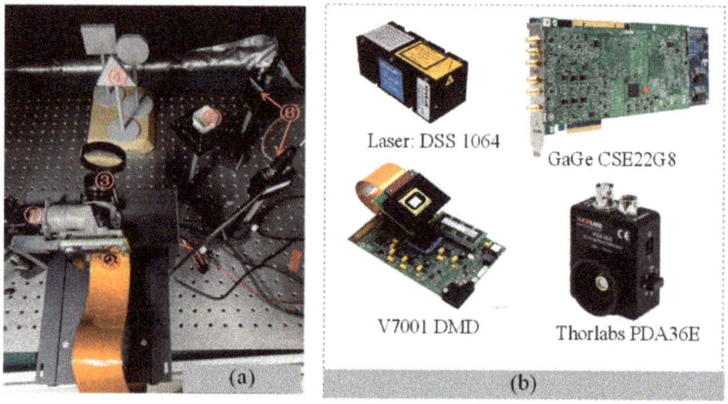

Figure 3. Indoor experimental setup. (**a**) The structure of the experiment; (**b**) the main components used in the indoor experiments.

Taking into account the actual working environment, the typical walking distance of the UAVs from the beginning to the detection is 100 m, as shown in Figure 4. The arch height of the railway tunnel in China is different according to the actual situation; for example, the height from the rail surface to the arch of the railway tunnel of the Guiguan Line is 8.68 m. Meanwhile, according to the Design Code for High-Speed Railway Tunnels, it is required to be no less than 8 m, and 10 m is taken for the calculation. Therefore, the covering distance of the system prototype should be no less than 100.5 m. In order to complete the laboratory experiment, the system was integrated, as shown in Figure 3a, and the system test environment is shown in Figure 3b.

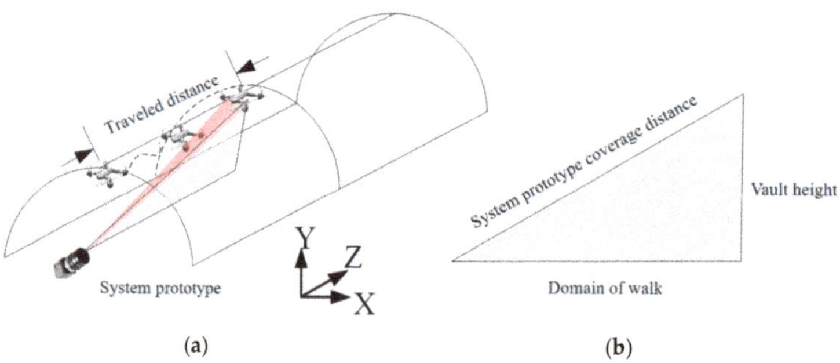

Figure 4. Schematic diagram of the working distance of the inner wall of the robot tunnel. (**a**) Working distance diagram; (**b**) range of coverage.

We carried out an infield experiment, in which the experimental equipment was placed inside the house; the specific location and system prototype are shown in Figure 5a,b. Two cables were placed 100 m away from the system; the test environment is shown in Figure 5c. The diameter of the cable was 0.5 cm, as shown in Figure 5d.

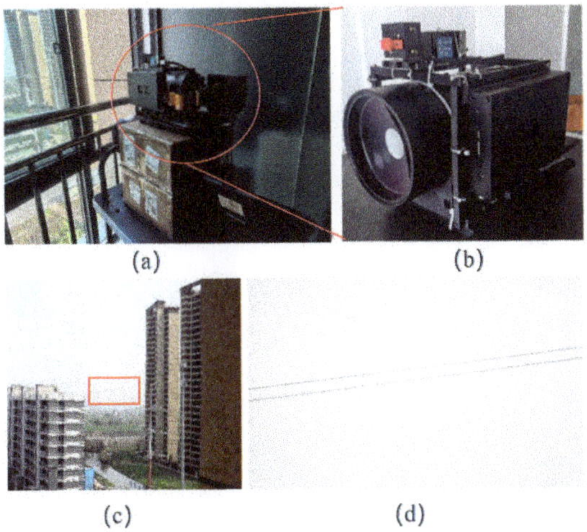

Figure 5. Setup of the infield experiment. (**a**) The location of the laboratory equipment; (**b**) enlarged view of the experimental setup; (**c**) the environment of the target object; (**d**) target object.

The prototype, as depicted in Figure 6c, integrates a 120 mm aperture catadioptric telescope for long-range single-pixel imaging of 100 m, enhancing the ability to detect small targets at a distance. Given the harsh conditions in a tunnel, such as dim lighting and the presence of dust, traditional imaging methods struggle to produce clear images. As shown in Figure 6b, even with a mobile phone flash at approximately 5 m, the visibility is significantly limited, and without illumination, as seen in Figure 6d, the cables are nearly indistinguishable to the naked eye. The core components of the rest of the system, including the pulse laser, DMD, and single-pixel detectors, are consistent with those used in our indoor experiments, ensuring a reliable baseline for comparison.

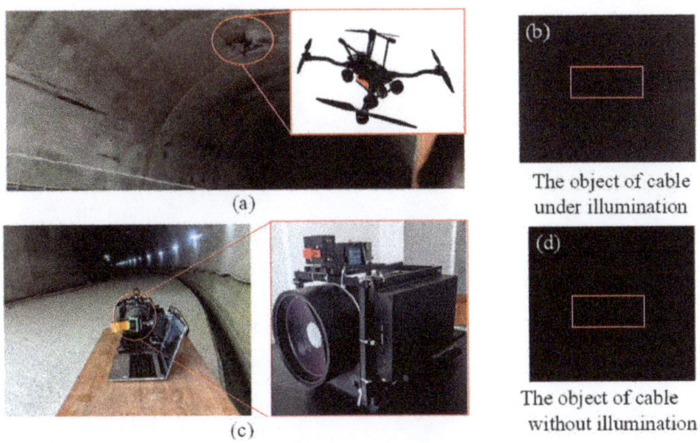

Figure 6. Field experimental setup. (**a**) Specialized robot under working conditions. (**b**,**d**) are the target of cable under illumination and without illumination. (**c**) The prototype is displaced at the tunnel entrance.

3.2. Results and Discussion

Our experiments aimed to validate the performance of the proposed eye-inspired variable-resolution single-pixel imaging technique under controlled indoor conditions and real-world field scenarios.

3.2.1. Data Compression Performance Verification (Indoor Experiments)

To compare the imaging effects, we used three basic geometric shapes: an equilateral triangle, a circle, and a square, each with a side length or diameter of 10 cm, placed at intervals of 8 cm and 14 cm from each other. The triangle was positioned 60 cm from the light source, covering the entire FOV, which was defined as the region of interest (ROI). We captured data with 2500 samples; the imaging results are shown in Figure 7, where (a) represents the reconstructed intensity maps at different distances, (b) shows the distance maps in different colors, and (c) presents a three-dimensional view.

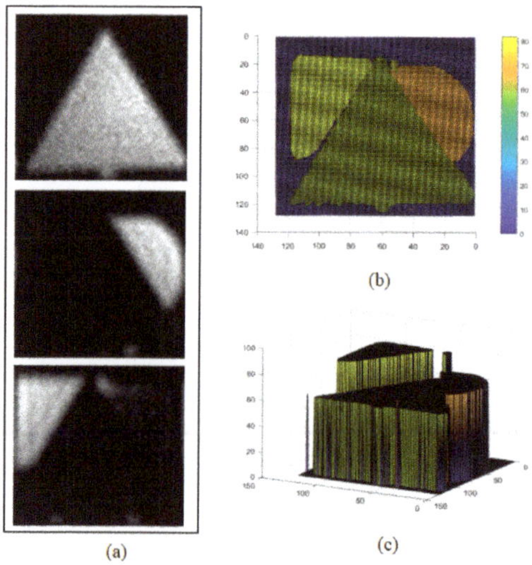

Figure 7. Traditional uniform-resolution random speckle 3D imaging results. (**a**) Reconstructed intensity map, (**b**) reconstructed distance map, and (**c**) 3D view.

Compared to the traditional uniform random speckle pattern reconstruction methods, our experiment applied a variable-resolution sampling technique, reducing the number of samples by half. This improvement demonstrates an increase in the reconstruction efficiency while maintaining the image quality, highlighting the potential of optimizing the reconstruction process in GI technology by adjusting the sampling strategy. When the distance between the triangular target and light source was increased to 100 cm, occupying only a part of the imaging FOV, we optimized the imaging effect by selecting the central area of the FOV as the ROI. By mimicking the retinal structure of the human eye, we generated a high-resolution speckle pattern in this area. The central area, with a radius of 42 pixels, was surrounded by five concentric circular rings, each containing 72 pixels, totaling 4432 pixels. The central region was filled with variable-resolution speckles, similar to the human eye, while the peripheral area was filled with random speckles. Even when the sample number was reduced to 1500, which was only 0.1 of the traditional uniform speckle sampling rates, the reconstructed image still achieved good quality, as shown in Figure 8. This result confirms that with an optimized speckle pattern, clear imaging results can still be obtained even at a reduced sampling ratio, thereby effectively verifying the characteristic of variable resolution in compressing redundant data.

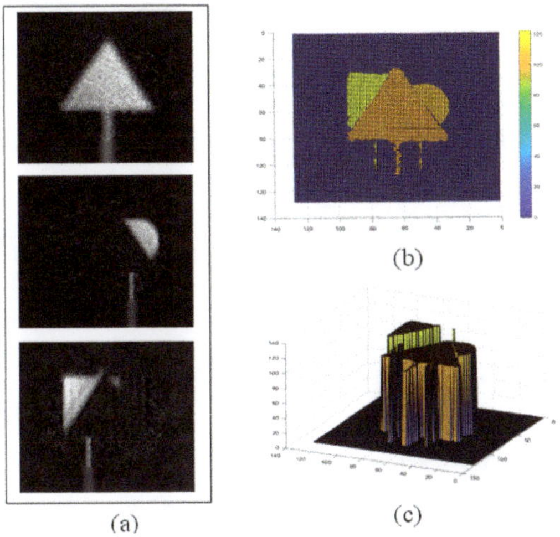

Figure 8. Three-dimensional imaging results using eye-inspired variable-resolution speckle method. (**a**) Reconstructed intensity map, (**b**) reconstructed distance map, and (**c**) 3D view.

3.2.2. Fine Target Imaging Capability Verification

Two cables were placed 100 m away from the system to image them, and the reconstruction quality was compared under different association times, as shown in Figure 9, which shows the system imaging SNR with the system sampling rate ranging from 10% to 100%. It can be seen from the figure that even if the sampling rate is reduced to 10%, the resulting image is still able to maintain a signal-to-noise ratio of up to 35 dB. This shows that even under the condition of a low sampling rate, the system constructed in this paper can still obtain high-quality images, so as to support the UAVs and help them to more accurately perceive the environment and avoid obstacles.

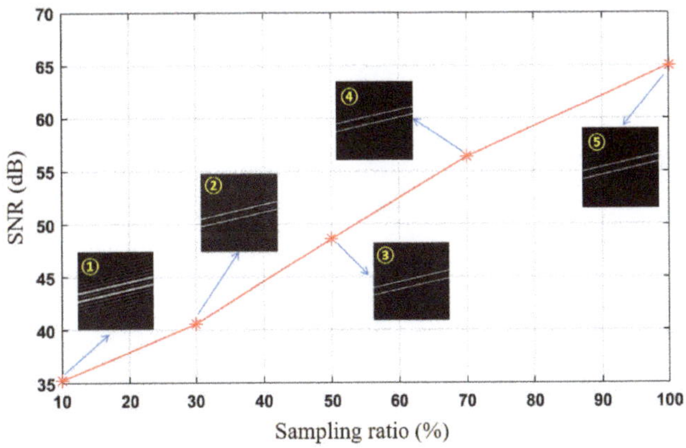

Figure 9. SNRs under different sampling ratios. The sampling ratios from ① to ⑤ were 10, 30, 50, 70, and 100%, respectively.

After verifying the feasibility of the system prototype at the simulated site, the system prototype was deployed to the actual application environment, and the site was selected as a section of closed tunnel in Xiongqin (Xiongan to Qinzhou), as shown in Figure 10. The goal is still to image the wire 100 m away from the system prototype; the diameter of the wire is 0.5 cm. The difference is that the tunnel environment is relatively dim, and because it is still in the construction stage, there is still a lot of dust inside, such test conditions are designed to simulate the challenges that may be encountered in practical applications, focusing on the imaging ability of small targets at a distance. Thus, the imaging advantage of this system in a complex environment is demonstrated.

Figure 10. A section of closed tunnel in Xiongqin (Xiongan to Qinzhou).

From the experimental environment, it can also be seen that the bad environmental conditions put forward higher requirements for imaging technology. Factors such as lighting and dust in real application scenarios have a significant negative impact on image quality compared to simulation scenarios. In such an environment, it is often difficult for conventional imaging techniques to capture clear and accurate images, resulting in a limited ability to identify and analyze distant targets, as shown in Figure 6b,d.

As emphasized in the experimental preparation, traditional imaging methods struggle to discern the cable targets. Under the same test conditions, our prototype successfully imaged distant wires, demonstrating the ability to maintain a high imaging quality in environments that are both dimly lit and have poor clarity, as shown in Figure 11. The corresponding imaging results at sampling ratios of 10%, 30%, 50%, 70%, and 100% are presented, with quantitative results showing SNR ratios of 13.5 dB, 16.1 dB, 19.6 dB, 24.2 dB, and 27.7 dB, respectively. This experiment revealed that under low-light and dusty conditions, our method is more conducive to capturing images of fine targets.

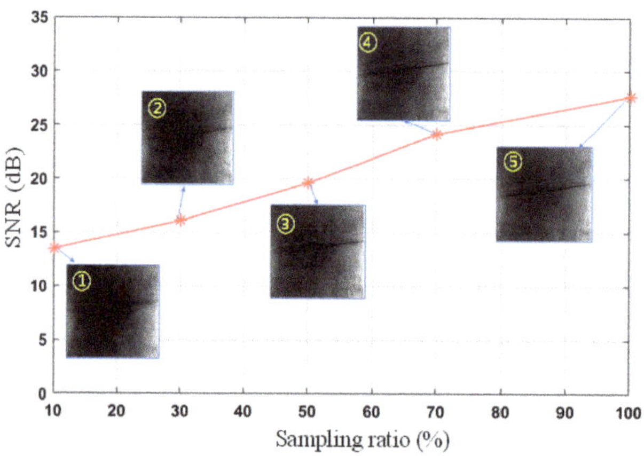

Figure 11. SNRs under different sampling ratios. The sampling ratios from ① to ⑤ were 10, 30, 50, 70, and 100%, respectively.

By comparing the imaging results under different sampling ratios, we found that under low sampling ratio conditions, single-pixel imaging technology combined with variable-resolution sampling and lateral inhibition characteristics could provide higher SNR ratios and clearer images. For instance, in an actual tunnel environment, the imaging result at a 10% sampling ratio had an SNR ratio of 13.5 dB, whereas the result at a 100% sampling ratio had a ratio of 27.7 dB. The experiments demonstrated that, even under the dim conditions inside the tunnel, the prototype system could successfully reconstruct recognizable images, which is crucial for guiding robots to avoid obstacles. This capability not only verified the robustness of the system but also showed its good environmental adaptability in maintaining relatively clear imaging under extreme conditions.

4. Conclusions and Future Works

The experimental results validated the effectiveness of our proposed eye-inspired variable-resolution single-pixel imaging technique for enhancing the environmental perception in special unmanned vehicles (UAVs). This innovative approach, tested extensively in both controlled indoor settings and real-world tunnel scenarios, demonstrated its potential for practical applications. In particular, in GPS-denied tunnel environments with dim lighting and dust, the technique showed a significant improvement in the efficiency and quality of image reconstruction for fine targets. The integration of lateral inhibition was also successful in reducing noise and artifacts, resulting in cleaner image reconstruction.

In future work, the primary focus will be on conducting more comprehensive field tests to confirm the robustness of the system under diverse environmental conditions. There is a clear need to develop advanced image-processing algorithms to achieve real-time imaging capabilities and to integrate this imaging system with autonomous robots or unmanned aerial vehicles (UAVs) for automated inspection tasks. Aligning technology with industry standards and regulatory requirements is also a priority for expediting commercial deployment. Our goal is to develop this technology into a reliable tool that can improve safety and efficiency in tunnel inspection and other applicable areas. By providing clearer and more accurate images than traditional imaging techniques, we hope to help inspectors spot potential problems faster so they can take preventive measures to avoid possible accidents and delays." This will mark an important advance in imaging technology, providing more efficient and safer solutions for a variety of inspection and monitoring tasks.

Author Contributions: Conceptualization, B.H. and J.C.; methodology, B.H.; software, Q.Z.; validation, Q.Z., B.H. and K.W.; formal analysis, J.C.; investigation, Y.S.; resources, Y.S.; data curation, Q.H.; writing—original draft preparation, M.S.; writing—review and editing, M.S.; visualization, Q.Z.; supervision, J.C.; project administration, K.W.; funding acquisition, Q.H. All authors have read and agreed to the published version of the manuscript.

Funding: This work was supported by Beijing Nature Science Foundation of China (4232014), National Natural Science Foundation of China (62275022).

Institutional Review Board Statement: Not applicable.

Data Availability Statement: The data presented in this study are available upon request from the corresponding author.

Conflicts of Interest: Kexin Wang was employed by the company The China Railway 12th Bureau Group Company, Ltd. The remaining authors declare that the research was conducted in the absence of any commercial or financial relationships that could be construed as a potential conflict of interest.

References

1. Chang, Y.; Cheng, Y.; Manzoor, U.; Murray, J. A review of UAV autonomous navigation in GPS-denied environments. *Robot. Auton. Syst.* **2023**, *170*, 104533. [CrossRef]
2. Zhang, K.; Cao, J.; Zhang, F.H.; Jiang, Y.; Cheng, Y.; Wang, D.; Hao, Q. Fourier Single-pixel Imaging Based on Lateral Inhibition for Low-contrast Scenes. *IEEE Photonics J.* **2019**, *11*, 6901711. [CrossRef]

3. Li, H.; Liu, Z.; Lyu, Y.; Wu, F. Multimodal Image Registration for GPS-denied UAV Navigation Based on Disentangled Representations. In Proceedings of the 2023 IEEE International Conference on Robotics and Automation (ICRA), London, UK, 29 May–2 June 2023.
4. Chen, Y.; Jiang, J. An Oblique-Robust Absolute Visual Localization Method for GPS-Denied UAV With Satellite Imagery. *IEEE Trans. Geosci. Remote Sens.* **2024**, *62*, 5601713. [CrossRef]
5. Wang, F.; Wang, C.; Chen, M.; Gong, W.; Zhang, Y.; Han, S.; Situ, G. Far-field super-resolution ghost imaging with a deep neural network constraint. *Light Sci. Appl.* **2022**, *11*, 1. [CrossRef]
6. Baris, I.E.; Jeffrey, H.S. Ghost imaging: From quantum to classical to computational. *Adv. Opt. Photon.* **2010**, *2*, 405–450.
7. Shi, M.; Cao, J.; Cui, H.; Zhou, C.; Zhao, T. Advances in Ghost Imaging of Moving Targets: A Review. *Biomimetics* **2023**, *8*, 435. [CrossRef] [PubMed]
8. Moodley, C.; Forbes, A. Super-resolved quantum ghost imaging. *Sci. Rep.* **2022**, *12*, 10346. [CrossRef] [PubMed]
9. Phillips, D.B.; Sun, M.-J.; Taylor, J.M.; Edgar, M.P.; Barnett, S.M.; Gibson, G.M.; Padgett, M.J. Adaptive foveated single-pixel imaging with dynamic supersampling. *Sci. Adv.* **2017**, *3*, e1601782. [CrossRef] [PubMed]
10. Lyu, M.; Wang, W.; Wang, H.; Wang, H.; Li, G.; Chen, N.; Situ, G. Deep-learning-based ghost imaging. *Sci. Rep.* **2017**, *7*, 17865. [CrossRef] [PubMed]
11. Shapiro, J.H.; Boyd, R.W. The physics of ghost imaging. *Quantum Inf. Process.* **2012**, *11*, 949–993. [CrossRef]
12. Li, W.; Hu, X.; Wu, J.; Fan, K.; Chen, B.; Zhang, C.; Hu, W.; Cao, X.; Jin, B.; Lu, Y.; et al. Dual-color terahertz spatial light modulator for single-pixel imaging. *Light Sci. Appl.* **2022**, *11*, 191. [CrossRef] [PubMed]
13. Shapiro, J.H. Computational ghost imaging. *Phys. Rev. A* **2008**, *78*, 061802. [CrossRef]
14. Jiang, H.; Zhu, S.; Zhao, H.; Xu, B.; Li, X. Adaptive regional single-pixel imaging based on the Fourier slice theorem. *Opt. Express* **2017**, *25*, 15118–15130. [CrossRef]
15. Paniate, A.; Massaro, G.; Avella, A.; Meda, A.; Pepe, F.; Genovese, M.; D'Angelo, M.; Ruo-Berchera, I. Light-field ghost imaging. *Phys. Rev. Appl.* **2024**, *21*, 024032. [CrossRef]
16. Edgar, M.P.; Gibson, G.M.; Padgett, M.J. Principles and prospects for single-pixel imaging. *Nat. Photonics* **2019**, *13*, 13–20. [CrossRef]
17. Ryczkowski, P.; Barbier, M.; Friberg, A.T.; Dudley, J.; Genty, G. Ghost imaging in the time domain. *Nat. Photonics* **2016**, *10*, 167–170. [CrossRef]
18. Vaz, P.G.; Amaral, D.; Requicha Ferreira, L.F.; Morgado, M.; Cardoso, J. Image quality of compressive single-pixel imaging using different Hadamard orderings. *Opt. Express* **2020**, *28*, 11666–11681. [CrossRef] [PubMed]
19. Han, S.; Yu, H.; Shen, X.; Liu, H.; Gong, W.; Liu, Z. A Review of Ghost Imaging via Sparsity Constraints. *Appl. Sci.* **2018**, *8*, 1379. [CrossRef]
20. He, R.; Weng, Z.; Zhang, Y.; Qin, C.; Zhang, J.; Chen, Q.; Zhang, W. Adaptive Fourier single pixel imaging based on the radial correlation in the Fourier domain. *Opt. Express* **2021**, *29*, 36021–36037. [CrossRef]
21. Wang, G.; Zheng, H.; Tang, Z.; Zhou, Y.; Chen, H.; Liu, J.; He, Y.; Yuan, Y.; Li, F.; Xu, Z. All-Optical Naked-Eye Ghost Imaging. *Sci. Rep.* **2020**, *10*, 2493. [CrossRef] [PubMed]
22. Khakimov, R.I.; Henson, B.M.; Shin, D.K.; Hodgman, S.S.; Dall, R.G.; Baldwin, K.G.H.; Truscott, A.G. Ghost imaging with atoms. *Nature* **2016**, *540*, 100–103. [CrossRef]
23. Hao, Q.; Tao, Y.; Cao, J.; Tang, M.; Cheng, Y.; Zhou, D.; Ning, Y.; Bao, C.; Cui, H. Retina-like Imaging and Its Applications: A Brief Review. *Appl. Sci.* **2021**, *11*, 7058. [CrossRef]
24. Graydon, O. Retina-like single-pixel camera. *Nat. Photonics* **2017**, *11*, 335. [CrossRef]
25. Orange Kedem, R.; Opatovski, N.; Xiao, D.; Ferdman, B.; Alalouf, O.; Kumar Pal, S.; Wang, Z.; von der Emde, H.; Weber, M.; Sahl, S.J.; et al. Near index matching enables solid diffractive optical element fabrication via additive manufacturing. *Light Sci. Appl.* **2023**, *12*, 222. [CrossRef]
26. Etchepareborda, P.; Moulet, M.-H.; Melon, M. Random laser speckle pattern projection for non-contact vibration measurements using a single high-speed camera. *Mech. Syst. Signal Process.* **2021**, *158*, 107719. [CrossRef]
27. Chen, C.; Li, Z.; Fu, L. Perovskite photodetector-based single pixel color camera for artificial vision. *Light Sci. Appl.* **2023**, *12*, 77. [CrossRef] [PubMed]

biomimetics

Article

Visual System Inspired Algorithm for Enhanced Visibility in Coronary Angiograms (VIAEVCA)

Hedva Spitzer [1,*], Yosef Shai Kashi [1], Morris Mosseri [2,3] and Jacob Erel [4]

1 School of Electrical Engineering, Iby and Aladar Fleischman Faculty of Engineering, Tel-Aviv University, Tel-Aviv 69978, Israel; shaikashi@gmail.com
2 Sackler Faculty of Medicine, Tel-Aviv University, Tel-Aviv 69978, Israel
3 Cardiology Department, Meir Medical Center, Kfar-Saba 95847, Israel
4 Nuclear Cardiology Unit and CCT Service, Meir Medical Center, Kfar-Saba 95847, Israel
* Correspondence: hedva@tauex.tau.ac.il

Abstract: Numerous efforts have been invested in previous algorithms to expose and enhance blood vessel (BV) visibility derived from clinical coronary angiography (CAG) procedures, such as noise reduction, segmentation, and background subtraction. Yet, the visibility of the BVs and their luminal content, particularly the small ones, is still limited. We propose a novel visibility enhancement algorithm, whose main body is inspired by a line completion mechanism of the visual system, i.e., lateral interactions. It facilitates the enhancement of the BVs along with simultaneous noise reduction. In addition, we developed a specific algorithm component that allows better visibility of small BVs and the various CAG tools utilized during the procedure. It is accomplished by enhancing the BVs' fine resolutions, located in the coarse resolutions at the BV zone. The visibility of the most significant clinical features during the CAG procedure was evaluated and qualitatively compared by the consensus of two cardiologists (MM and JE) to the algorithm's results. These included the visibility of the whole frame, the coronary BVs as well as the small ones, the main obstructive lesions within the BVs, and the various angiography interventional tools utilized during the procedure. The algorithm succeeded in producing better visibility of all these features, even under low-contrast or low-radiation conditions. Despite its major advantages, the algorithm also caused the appearance of disturbing vertebral and bony artifacts, which could somewhat lower diagnostic accuracy. Yet, viewing the processed images from multiple angles and not just from a single one and evaluating the cine mode usually overcomes this drawback. Thus, our novel algorithm potentially leads to a better clinical diagnosis, improved procedural capabilities, and a successful outcome.

Keywords: blood vessel enhancement; coronary angiography; lateral interaction algorithm

1. Introduction

Coronary artery disease (CAD) is a major global cause of death [1], caused by atherosclerotic plaque buildup. It may cause gradual narrowing of the inner side of arteries over time or abrupt rupture and clotting that end up with partial or total blockage of the blood flow. The current gold standard for the detection and invasive treatment of CAD is conventional X-ray CAG [2]. A contrast agent is injected during the acquisition of CAG to improve the visibility of the BV and their blood flow. A CAG unavoidably exposes the patients to a limited controlled dose of harmful X-ray radiation and toxic contrast agents [3].

The capabilities to enhance CAG visibility post-acquisition are based on computational methods that modify brightness, contrast, and background by various algorithms. The quality of CAG images still suffers from low contrast (mainly due to dye clearance and low-radiation fluoroscopy runs), non-uniform illumination issues, and the presence of other body organs, which makes the diagnosis and medical interventional procedures a challenging task. A comprehensive algorithm that can simultaneously enhance the visibility of all image components, without losing details and even revealing important clinical information, is therefore, clearly needed.

Improved CAG visibility has the potential to help also in selecting the required equipment for the catheterization procedure. It supports smooth and faster navigation with the various catheters, especially with the thin wire inserted into the coronary blood vessel (CBV), which is commonly used during invasive treatment. This allows accurate deployment of balloons and stents over it, precisely at the right locations. A successful procedure includes many additional tools and elements, which are all dependent on visibility, such as placing the tip of the catheter at the proper angle in the orifice of the vessel, avoiding damage to the delicate endothelial tissue when the dye is injected at high pressure, or when entering and navigating with the tip of the wire inside the vessel and later deploying a stent over it. An additional advantage is avoiding unnecessary invasive treatment when identifying good flow through collaterals that bypass the obstructive lesion or that originate from other CBVs and supply sufficient blood to the same viable myocardial territory. Previous algorithms did not relate to this type of CAG enhancement and visibility.

Over the years, numerous studies have been developed for the segmentation and enhancement of BVs applied to different anatomical areas and imaging modalities [4,5]. These classical algorithms dealt mainly with these challenges through second derivative algorithms (i.e., Hessian algorithms) [6] and other algorithms that tried to improve the dynamic range of the image in order to solve illumination problems, such as contrast limited adaptive histogram equalization (CLAHE) [7,8], And additional study suggested morphological filters [9].

Hessian-based methods have commonly been applied for the purposes of blood vessel segmentation and enhancement in CAG images [10–13]. Although these studies obtained good segmentation of the main coronary arteries, only a small repertoire of the successful segmentation of small BVs was obtained. Moreover, these Hessian derivative algorithms also caused enhanced noise in the image [14]. Truc and his colleagues (2009) tried to reduce the noise by using a Hessian algorithm and applying wedge filters, as preprocessing components.

Some of these publications added additional factors that enhance and improve the blood vessel's appearance, such as adding the G-L differential kernel [15], which yielded better results than, for example, the classical CLAHE algorithm. Results of CLAHE algorithms have appeared many times with exaggerated results and with additional noise [16].

Few computational algorithms have been developed to reduce the amount of toxic X-ray radiation and contrast agent dose while minimizing image quality degradation [1,17,18]. For enhancing the visibility at low doses, different methods have been used, such as decreasing the noise through principal component analysis (PCA) [17], adding the isolated vessel layer to the low-dose image [18], and using a convolutional neural network (CNN) algorithm [1]. Exposure of additional BVs under the condition of low contrast was shown by using fractional differential kernels [19] and additional classical methods, such as Hessian-based methods [15] or lateral inhibition-based models.

Due to the substantial research efforts dedicated to improving CAG image quality issues at standard radiation, our study mainly refers to the most recent and relevant studies. Those studies have been developed in order to overcome several image visibility drawbacks, such as non-uniform illumination, artifacts and noise, low contrast, and interference of vertebral objects [15,18,20,21].

A group of other recent studies challenged the efficient segmentation of BVs through background subtraction algorithms. These studies aimed to isolate the BVs, in still or in video images, from the bony "artifacts" of the spinal cord and additional organs. The purpose of this approach is to allow the observation of BVs without "disturbing" vertebral bodies, such as bones and the spinal cord. These studies computationally separate the CAG image into two [20,22] and sometimes three different layers [18,21], namely, vessels, breathing, and background (spine and hard tissues) layers. Since the spatial information by itself cannot separate efficiently the three layers, most of the subtraction operations used mutual information of spatial and temporal aspects. Most of these algorithms utilized the known Robust Principal Component Analysis (RPCA) method [18,20,21]. Such a method enables the separation of the image into a matrix that is decomposed into a low-rank and sparse matrix, which represents the background and vessel layers. These subtraction studies [18,20,21] succeeded in separating the background from the BV image, but none of them included the small BVs in their BV layer.

In recent years, CNNs have become a common method for coronary blood vessel segmentation [8,23–26]. Most of the studies on this topic did not include exposure of small blood vessels in their ground-truth image [8,23,26], and therefore, their segmentation results did not reveal exposure of small blood vessels. Only more recent papers, from 2021 [24,25], presented ground-truth images that supplied a larger repertoire of blood vessels, including the small blood vessels, to the U-net and Random Forests Classifier algorithms. Their results yielded better performance than the previous algorithms, but still, the visibility of small blood vessels was limited [25].

Although many studies and computational methods addressed the CAG image quality issues, it seems that solutions for the clear appearance of small blood vessels and visibility of the content inside the large blood vessels with their relevant catheterization apparatus are still required. Revealing these details and objects can be critical in allowing an efficient diagnosis. We present here a single algorithm that addresses both the visibility of BVs and the catheterization tools. It simultaneously reduces noise and illumination artifacts and can also enhance the visibility of BVs under the conditions of a low-contrast agent dose and reduced radiation. The algorithm is based on a new biologically inspired mechanism and a computational model (lateral facilitation) [27,28].

2. Proposed Algorithm

2.1. Rational of the Algorithm

In this study, we developed and applied different algorithm components for enhancing the visibility of CBVs and the various tools utilized during the CAG procedure. For this purpose, we first used a preliminary stage aimed at enhancing the visibility of the relevant region of interest (ROI) in the images, namely, the CBVs. For a successful diagnosis, cardiologists need to properly evaluate the whole CBV system, including the small vessels, and identify all significant obstructive lesions. Enhancing the visibility of the catheterization tools within the blood vessels is also important for successful invasive procedures. The algorithm was developed to overcome quantum noise, artifacts, non-uniform illuminations, and low contrast due to a low-radiation and/or low-contrast agent dose, which can all substantially decrease visibility.

The suggested algorithm is based on Multiscale Gabor filters having different orientations and scales (similar to what occurred with the different BVs). The main contribution of the algorithm to the BV enhancement is inspired by a visual system mechanism and its model [27]. This mechanism is known as lateral facilitation or lateral interactions, while the mechanism of contour integration is also included in the suggested model [27–29]. (2) The model suggests an additive facilitation sub-threshold signal, which is added to an area outside the classical receptive field but that is induced at the optimal relevant orientation of the classical receptive field. This sub-threshold response may be enhanced to become suprathreshold by the additive signal of the neighboring receptive fields (RFs) (in the case that a similar orientation preference is shared between the two adjacent receptive fields). The possibility that the additive signal will cause a super-threshold response and fulfill contour integration can be regarded as determined by an educated guess signal (additive signal). The line and texture completion of the blood vessels enables us to enhance their structure at the expense of part of the noise in an image.

2.2. Overview

Our algorithm for blood vessel visibility enhancement is based on several building blocks, which are summarized in the schematic diagram of Figure 1. Each component is represented by a different background color.

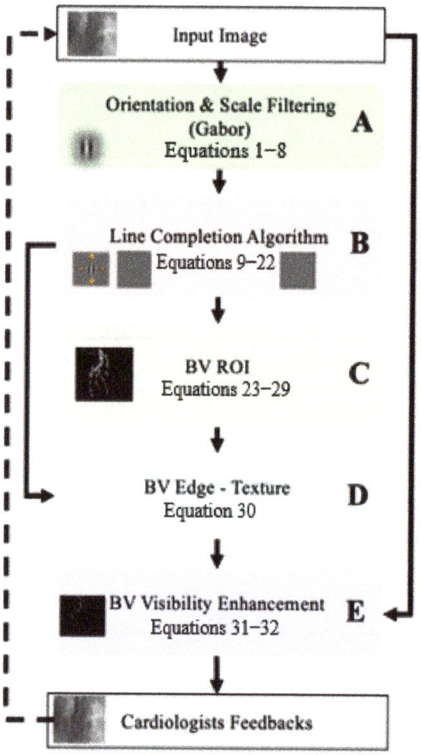

Figure 1. The schematic building block of the main components of the VIAEVCA algorithm. Each colored block represents a different algorithm component (with reference to the appropriate equation numbers in the model). The same code of block colors is given for the different algorithm components and will be used for the rest of the schematic building blocks presented along the algorithm components description below.

The algorithm's first stage (Figure 1A, green block) performs Gabor filtering in several orientations and scales, which is similar to what is performed with the simple cells' receptive fields of the visual system. The second stage (Figure 1B, pink block) performs the line completion algorithm, which is based on computational models aimed to supply mechanisms and models to visual phenomena, such as lateral facilitation, and collinear interaction stages. The third stage (Figure 1C yellow block) performs the extraction of the blood vessel's region of interest (ROI). Then, in the fourth stage (Figure 1D cyan block), the ROI result is used to extract the blood vessels from the image of the line completion results (Figure 1E pink block). The resulting image contains the blood vessel edges' texture (which has been obtained from the line completion) but only at the zone of the blood vessel locations (BV ROI). The purple block represents blood vessel visibility enhancement after the integration of the blood vessel edge texture with the original image (DC of the image). The algorithm results (purple block) have been presented to the cardiologists for estimation and feedback. The additional stage of algorithm corrections, according to the cardiologists' feedback, is the final estimations, which are presented in the Section 5.

2.3. Orientation and Scale Filtering (Gabor)

The first stage of the algorithm is intended to extract edges, which are built from different scales (j) and different orientations (θ). The edge image, $s(x,y,j,\theta)$, is obtained through the convolution of Gabor kernels $G_{\hat{\theta}}(x,y)$, Equation (2), while using a Gaussian Image Pyramid deconstruction of the input image $I(x,y)$, which is built from different scale images, $I^j(x,y)$. The edge image, thus, is built of edges with different orientations and different spatial resolutions. Equation (1) represents the calculation of the Gabor kernels at different orientations:

$$G_{\hat{\theta}}(x,y) = \frac{1}{g} e^{-\frac{(x-x_0)^2+(y-y_0)^2}{\sigma^2}} \cdot \cos\left(\frac{2\pi}{\lambda} \cdot (x-x_0) \cdot \cos(\hat{\theta}) + (y-y_0) \cdot \sin(\hat{\theta})\right) \tag{1}$$

where the frequency ($\frac{1}{\lambda}$) and the orientation (θ) of the spatial filter are determined by the frequency and orientation at each spatial location (x,y). σ is the standard deviation of the Gaussian envelope. (x_0,y_0) is the spatial location of the center of the Gabor filter. g is chosen as the normalization factor.

To estimate the orientation feature of images' edges (stimulus), we first convolved the Gabor kernel at different orientation values:

$$s(x,y,j,\theta) = \begin{cases} \sum\limits_n \sum\limits_m I^j(m,n) \cdot G_\theta(x-m, y-n) & x,y \ at \ RFs \ center \\ 0 & O.W \end{cases} \tag{2}$$

where $I^j(x,y)$ is a Gaussian Image Pyramid deconstruction of $I(x,y)$, and j represents the spatial scale.

After the edge (sparse) image $s(x,y,j,\theta)$ is obtained, we determine the optimal edge orientation at each location (x, y), separately for each scale (j). Therefore, in Equation (3) $S(x,y,j)$, Figure 2C is chosen as the strongest convolution response (Equation (2)).

$$S(x,y,j) = \arg \max_{\theta \in [0,90]} (s(x,y,j,\theta)) \tag{3}$$

Due to the Gabor structure, which contains positive and negative areas, the optimal orientation edge image $S(x,y,j)$ contains both the strongest response in the positive and the negative value ranges. These two maximum responses are considered separately, as shown in Equations (4) and (5).

$$c_P(x,y,j) = \max(S(x,y,j), 0) \tag{4}$$

$$c_N(x,y,j) = \max(-S(x,y,j), 0) \tag{5}$$

where c_P and c_N represent the largest positive and negative responses, respectively.

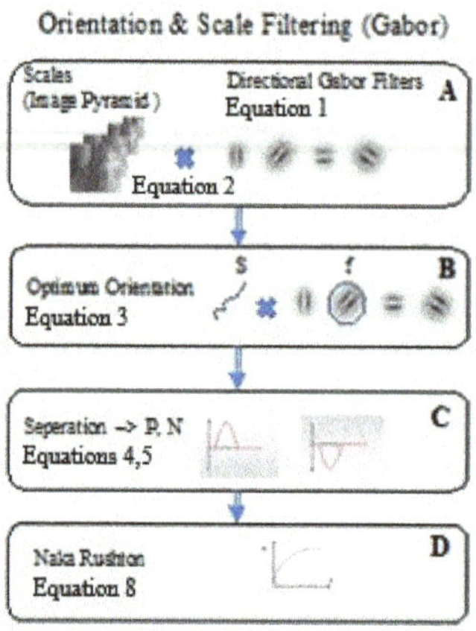

Figure 2. Schematic block diagram of the initial part of the algorithm that computes the different Gabor receptive fields of orientations and spatial resolutions. (**A**) Direclioral Gabor Filters. (**B**) The optimal orientation and scale filters are then computed. (**C**) The negative and positive components are separated. The small plots on the right and left sides illustrate the mathematical operations. (**D**) The Naka–Rushton operation.

For the sake of simplicity, from this point and further on, all the equations refer to P and N values as one group, although the computations are independent. The positive and negative components, until the components are separated, are recombined at the final stage of the algorithm. $c(x,y,j)$ is, therefore, used to represent both $c_P(x,y,j)$ and $c_N(x,y,j)$.

To decrease the amount of noise in the image, all intensity responses below 5% are considered here as noise (Equation (6)). The threshold, *thr*, is consequently calculated as the 5% percentile, across all the maxima response values and across all the spatial locations in the edge image $c(x,y,j)$. Equation (6) is shown below.

$$c(x,y,j) = (c(x,y,j) > thr) \cdot c(x,y,j) \quad (6)$$

where

$$thr = \pm \max|c(x,y,j)| \cdot 0.95 \quad (7)$$

Following the lateral facilitation model [27], we applied the Naka–Rushton Equation $R_{RF}(x,y)_j$ (Equation (8)), Figure 2D, which was suggested as a response to the V1 simple cell's responses to the edge images $c(x,y,j)$. This operation enables us to enhance the gain response in reference to a specific stimulus intensity range.

$$R_{RF}(c(x,y,j))_j = \frac{c(x,y,j)^n}{c(x,y,j)^n + \sigma_{NR}^n} \quad (8)$$

where $R_{RF}(x,y)_j$ is the Naka–Rushton response of a cRF cell at location (x, y) and scale j. For the sake of simplicity, the NR parameters n and σ_{NR} are considered as constants.

2.4. Line Completion Algorithm

Figure 3 shows the schematic diagram of the line completion algorithm's component.

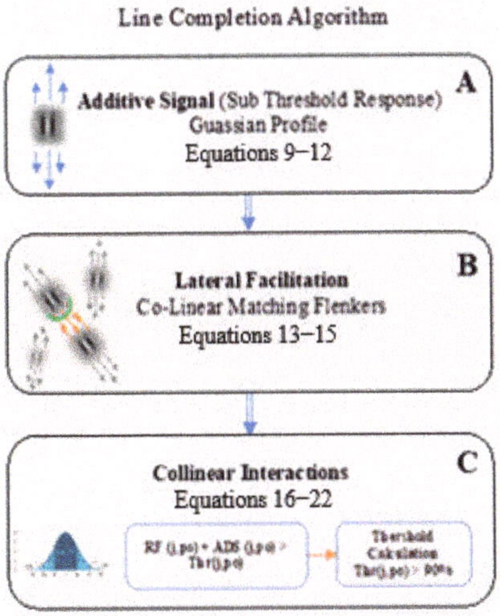

Figure 3. Schematic diagram of the line completion algorithm's component. (**A**) represents the additive signal, which refers to the extra-classical receptive field area that is located collinearly to the receptive field (the blue arrows). (**B**) illustrates the case that lateral facilitation is fulfilled when the additive signals of collinear responses (grey arrows) are overlapped (orange arrows). The green half-circle represents the weight function for the directional facilitation between adjacent receptive fields, through a Gaussian profile; see text. (**C**) illustrates the threshold response that was determined by the percentile response (blue Gaussian, **left**).

3. Lateral Facilitation

3.1. Additive Signal

Following the rationale of the algorithm (Section 2.1), the algorithm is presented in terms of the model of line completion, i.e., lateral facilitation [27]. An additive signal (*AdS*) is added beyond the response to the classically relevant receptive field (*RF*), as an educated guess response (Equation (15)). It is taken, therefore, as a sub-threshold response that is added to the *RF* optimal response, which has been yielded from the preferred appropriate orientation receptive field. Consequently, this response is calculated as a sub-threshold signal, which is spatially located along its relevant *RF*. It becomes suprathreshold only when there is an "agreement" with the neuronal mechanism of lateral interaction [28,29]. For the sake of simplicity, we present first the additive signal spatial decay profile, which is built from Gaussian functions (Equations (9)–(12)), and then the induced additive signal response, while also including its intensity profile and not only its spatial profile (Equation (13)); see Figure 2C.

Since the additive signal, *AdS* (Equation (9)), is added collinearly, it is expressed as a Gaussian mask located along the *RF* orientation, x-axis, which indicates the collinearity to the core of the receptive field. This additive signal, Ads, is added to the maximum responses of orientation and resolutions, $s(x, y)$, as sub-threshold signals.

$$AdS(x_i, y_i, \hat{\theta}\,) = \frac{1}{2\sigma_x} e^{\left(-\frac{1}{2}\left(\frac{x'}{\sigma_x}\right)\right)} \tag{9}$$

where x' is the collinear axis, $\hat{\theta}$ is the orientation of the optimal receptive field, and σ_x is defined as the designated number of pixels needed for line completion along the θ orientation.

To enable flexibility in the connectivity between collinearity additive signals, which is supposed to perform an efficient line completion, the model suggests a second axis Gaussian, y', (perpendicular to the *RF* length—x' axis), as a weighted function (WF) (Equation (11)). This component enables additivity between a neighbor's *RF* with also only similar orientations.

$$\mathrm{WF}(x_i, y_i, \hat{\theta}\,) = \frac{1}{2\sigma_y} e^{(-\frac{1}{2}(\frac{y'}{\sigma_y}))} \tag{10}$$

where σ_y is the weight function of the decay of the degree of the additive signal collinearity.

The final additive signal profile (AdS_f) is presented by a two-dimensional Gaussian function (Equation (12)).

$$AdS_f(x_i, y_i, \hat{\theta}\,) = AdS(x_i, y_i, \hat{\theta}\,) \times \mathrm{WF}(x_i, y_i, \hat{\theta}\,) = \frac{1}{2\sigma_x\sigma_y} e^{(-\frac{1}{2}(\frac{x'}{\sigma_x} + \frac{y'}{\sigma_y}))} \tag{11}$$

where $x' = x_i \cos(\hat{\theta}) + y_i sin(\hat{\theta})$, $y' = y_i \cos(\hat{\theta}) - x_i sin(\hat{\theta})$.

The additive signal, which includes its intensity and spatial profile is presented as follows:

$$RF_{AdS\theta,j}(x_i, y_i) = f\left(d^i, RF^i(x_i, y_i)_{\theta,j}\right) = \begin{cases} a \cdot R^i_{RF}(c(x_i, y_i))_{j,\theta} \times AdS_f(x_i, y_i, \theta) & d^i > r_{RF} \\ 0 & d^i \leq r_{RF} \end{cases} \tag{12}$$

where $R^i_{RF}(c(x_i, y_i))_{j,\theta}$ is the cRF response of the inducing *RF*. d^i is the distance from the center of the inducing *RF* to the edge of the receptive field and the distance of the additive signal, which includes the regions outside of the classical receptive field in the preferred orientation. r_{RF} is the distance between the center of the receptive field to its edges. a represents the strength of the induced *LF* (lateral facilitation).

3.2. Collinear Interactions

This stage of the model examines the possibility of the lateral facilitation operation at each image location; see Figure 2C. In other words, the model examines whether the sub-threshold additive signal, $RF_{AdS\theta,j}(x_i, y_i)$ (Equation (13)), at each specific *RF*, is changed to be super-threshold, in dependence on the responses at the neighborhood of the collinear neighborhood regions, $RF^{k(x_i, y_i)}_{AdS\ \ \theta,j}(x_{i+\cdot}, y_{i+\cdot})$. The model, therefore, examines whether there are collinear matching flankers that can overlap with their additive signals and, consequently, cause a super-threshold response. This additive signal interaction, thus, can fulfill the condition for the lateral facilitation operation (Equation (14)).

$$RF_{LF}(x_{i+\Delta}, y_{i+\Delta})_{\theta,j} = RF^i_{AdS\theta,j}(x_i, y_i) + RF^k_{AdS\theta,j}(x_{i+\Delta}, y_{i+\Delta}) \tag{13}$$

where RF_{AdS} is the receptive field additive signal at a specific location, while i and k are the indexes that present two different neighboring receptive fields, RF_{AdS}.

Until this stage, we considered only the facilitation obtained from the responses of adjacent collinear receptive fields, Equation (14). The model should also consider a possible response, R_{RF}, in case there is a stimulus located in the region of the overlapping summated receptive field with their additive signals, RF_{LF}. (The above consideration agrees with many experimental studies that have been completed on lateral facilitation and lateral integration (i.e., ref. [24])).

The lateral facilitation procedure, Equation (15), includes possible summation from all the receptive fields with their overlapping additive signals, along with the R_{RF}, located in the summation spatial area; see above.

$$LF(x,y)_{\theta,j} = R_{RF}(x,y) + \sum_i RF^i_{LF}(x_{i+\Delta}, y_{i+\Delta})_{\theta,j} \tag{14}$$

where $LF(x,y)_{\theta,j}$ represents the sum of the lateral facilitation signals at each *RF* location (x, y), along with $\mathrm{R_{RF}}(x,y)$; see Equation (9). Note that there is an additional stage of the threshold; see Equations (16) and (17).

The same logic of fulfilling the lateral facilitation is included for the whole image at all locations, through the summation presented in Equation (16). $LF(x,y)_{\theta,j}$ presents the last stage of the lateral facilitation procedure, for the whole image, but before an additional stage of thresholding, as presented in Equations (17) and (18).

$$LF(x,y)_{\theta,j} = c_{\theta,j}(x,y) + \sum_i RF^i_{LF}(x_{i+\Delta}, y_{i+\Delta})_{\theta,j} \tag{15}$$

where $LF(x,y)_{\theta,j}$ represents the sum of the lateral facilitation signals at each *RF* location (x, y), along a similar orientation θ and scale j. $c_{\theta,j}(x,y)$ represents the cRF response at location (x, y).

Following the previous section that described lateral facilitation, $LF(x,y)_{\theta,j}$, which is fulfilled only in the case that there is overlapping between adjacent additive signals of adjacent receptive fields, an additional threshold level, the thr_j stage, has been added, $LF\_thr(x,y)_{\theta,j}$; see Equation (17). The level of the threshold is determined according to percentile measure; see Equation (18). The aim of the thresholding stage, Equation (17), is to suppress *LF* responses, which are too low. This stage allows the algorithm a further reduction of the enhanced noises.

$$LF\_thr(x,y)_{\theta,j} = \max\left(0,\ LF(x,y)_{\theta,j}\right) \cdot \mathrm{H}(LF(x,y)_{\theta,j} - thr_j) \tag{16}$$

where H is a soft-threshold function.

$$thr_j = P^j_{\gamma_j\%}(LF(x,y)_{\theta,j}) \tag{17}$$

where $P^j_{\gamma_j}$ is the γ_j% percentile of the lateral facilitation response values, $LF(x,y)_{\theta,j}$, in Equation (18), across the whole image. γ_j is the percentile constant per scale j.

The algorithm components have been calculated separately for each response feature: polarities (P, N), orientations (θ), and scales (j) (see above). The following stages describe how the different components are combined to have final lateral facilitation responses while including all their features. The integration of the different components has been described separately for each algorithm's feature.

Polarities: The simulations have been treated separately for the positive and negative responses (Section 2.3; Equations (4) and (5)). However, our model omits writing each component twice, for the sake of simplicity in the previous sections.

We first start with two *LF* polarities, the negative $LF\_thr(x,y)^N{}_{\theta,j}$ and positive $LF\_thr(x,y)^P{}_{\theta,j}$.

$$LF\_thr(x,y)^O{}_{\theta,j} = LF\_thr(x,y)^P{}_{\theta,j} - LF\_thr(x,y)^N{}_{\theta,j} \tag{18}$$

Orientations: In the following stage, the algorithm sums up all the responses of the optimal orientations, which have been found for each Gabor Orientation, $LF\_thr(x,y)^O{}_{\theta,j}$ (Equation (2)).

During this summation process, we would like to also gain noise reduction, and therefore, we enhance the larger responses and suppress the smaller responses, by raising $LF\_thr(x,y)^{o}{}_{\theta,j}$ over the power of an integer m1 $\geq$ 1. To maintain the response sign (positive or negative response) of T_j, the calculation is separated into sign and absolute value components, in the following manner:

$$T_j = \sum_{\theta}\left[\left(\left|LF\_thr(x,y)^{O}{}_{\theta,j}\right|\right)^{m1} \cdot sign\left(LF\_thr(x,y)^{O}{}_{\theta,j}\right)\right] \tag{19}$$

To maintain the original *LF* intensity values range, while preserving the values signs and the enhanced contrasts, the algorithm applies an inverse operation to Equation (20):

$$T_j{}' = sign(T_j) \cdot \sqrt[m1]{|T_j|} \tag{20}$$

Scales: Equations (22) and (23) describe the operation of combining the collapsed scales. This is completed similarly as is completed for the recombination of the different orientations (Equations (20) and (21)). However, in this case, we had to consider the different scale image sizes. Before the scale recombination, each $T_j{}'$ (Equation (21)) is interpolated back into the size of the original image scale (see Section 4). The resized set of $T_j{}'$ responses are notated as $T_j{}''$.

$$T = \sum_{j}\left[\left(\frac{1}{\tau_j}\left|T_j{}''\right|\right)^{m2} \cdot \mathrm{sign}(T_j{}'')\right] \tag{21}$$

where m2 is constant (m2 $\geq$ 1), andτ_j is a scale-dependent constant, which is defined per each scale.

$$T_E = \gamma \cdot \mathrm{sign}(T) \cdot \sqrt[m2]{|T|} \tag{22}$$

where γ is a normalization constant.

3.3. BV ROI

The extraction stage of the region of interest (BV ROI) of the BV is designed to extract the locations of BVs, the catheterization apparatus, and the contrast agent (Figure 4). It is completed, thus, also to locate and enhance the catheterization apparatus, only at these spatial BV areas. The ROI estimate image is computed for each scale (j) separately, at the first stage (Figure 4A). The ROI estimate is a binary image, while the white pixels represent the ROI, and the black pixels represent the background. The binary image is calculated according to the compound's signal intensity values $T_j{}'(x,y)$ (Equation (23)), through the following stages. First, the histogram of each compound signal scale is computed, while each bin represents the probability of the range of intensity values of the compound responses $T_j\prime(x,y)$. The intensity threshold is calculated according to Otsu's method. Otsu's method chooses a threshold that minimizes the intra-class variance of the threshold black and white pixels, and therefore, provides a simple and compatible threshold measure that is appropriate for determining an estimate that is based on intensity responses. Equation (24) applies Otsu's threshold to create the binary image.

$$wb\,(x,y,j) = \begin{cases} 1 & if\ T_j'(x,y) > \mu_T(j) \\ 0 & else \end{cases} \tag{23}$$

where $\mu_T(j)$ is the Otsu threshold found for T_j'.

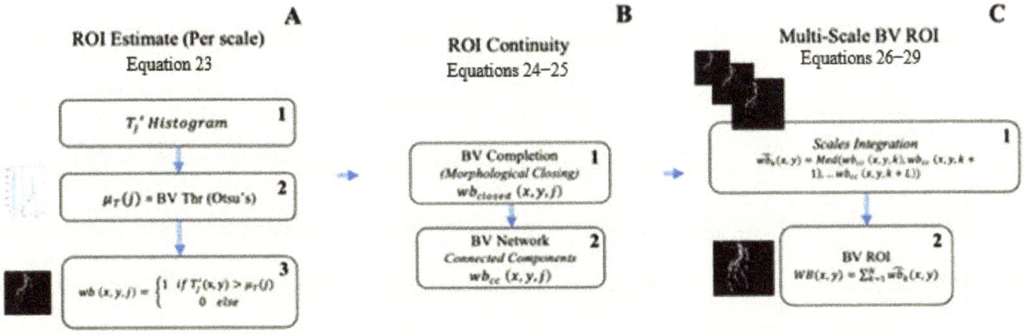

Figure 4. Schematic diagram of the blood vessel ROI detection, through three main stages. (**A**) illustrates the acquired binary ROI estimate (3), through the application of Otsu's threshold method on the *LF*(*j*) "image" (1–2) (left blue histogram). (**B**) represents the calculations of closing and connected operations. (**C**) represents the integration of the different response scales to compose the overall blood vessel's ROI (2).

The white pixels ($wb\ (x, y, j) = 1$) represent the spatial locations (*x*, *y*) of the BV ROI at the estimated image (Equation (24)), at each scale (*j*).

The additional ROI stage (Figure 4B), Equations (25) and (26), enables smooth continuity of the ROI region while selecting the major components in the image. The rationale of this stage is based on preserving the continuity of the physical BV tree, which could have been disrupted during the acquisition of the 2D CAG. Additional factors can interfere with achieving a non-ideal video image, which is needed for an optimal diagnosis. These factors are as follows: 1. Those that originate from deficient image quality due to limited X-ray or contrast agent doses, quantum noise, screening angle, non-uniform illumination, obstructions by different body organs, etc. 2. Disruptions that could have occurred from the threshold holding stages, at an earlier algorithm stage (Equation (24)). This threshold stage generates merely an estimate of the ROI; therefore, it may lead to some ROI areas that do not cross the threshold. This can lead to irregularities in the BV's structure, such as black holes or gaps in the binary images. Thus, a selection operation such as morphological closing (Equations (25) and (26)) seems appropriate. (It is worth noting that the usage of morphological closing on the binary image cannot create irrelevant or new information, since this image will only be used for ROI extraction).

The dilation operation of $wb\ (x, y, j)$ is obtained from the following:

$$wb_{dilated}\ (x, y, j) = \cup_{t \in B} wb(x, y, j)_t \tag{24}$$

where *B* is a binary structuring element and $wb(x, y, j)_t$ is the translation of wb by t. The closing operation of $wb\ (x, y, j)$ is obtained from the following:

$$wb_{closed}\ (x, y, j) = \cap_{t \in B} wb_{dilated}\ (x, y, j)_{-t} \tag{25}$$

where *B* is a binary structuring element with a radius r_j, and $wb_{dilated}\ (x, y, j)_{-t}$ is the translation of $wb_{dilated}\ (x, y, j)$ by $-t$.

The two preceding stages, the ROI estimate creation (Equation (24); Figure 4A) and the morphological closing (Equation (26); Figure 4B1)), which yielded a binary continuous BV tree, enabled us to use the connected-component labeling (CCL) algorithm, separately for each scale. This capacity of the CCL (Figure 4B2) to enhance the ROI was enabled after the creation of the largest connected tree, $wb_{closed}\ (x,y,j)$ (Equations (24)–(26)). The final stage of the CCL algorithm accounts for only the largest components in the image, $wb_{closed}\ (x,y,j)$, which are represented by $wb_{cc}\ (x,y,j)$.

Up to this stage, each ROI candidate (Equations (24)–(26)) was considered separately for each image scale, whereas the final ROI must include all the scales. The final stage of the ROI BV algorithm (Figure 4C), consequently, provides weighs function information (Figure 4C1; Equations (27) and (28)) from the different binary image scales $wb_{cc}\ (x,y,j)$, to enable the integration of various BV scales in a later stage (Figure 4C2, Equation (29)).

The main purpose of this part of the algorithm, as mentioned above, is to enhance the information inside the large BV that might contain information on additional smaller BVs, or on the catheterization tools. To perform this enhancement, we reorganize the histogram of the scale j partitions into new scale groups k, through different sizes of the interval (bins) and their degree of overlapping (bins' width in the histogram). This partition is completed by smoothing the median values of the intervals of the samples (bins) of the histogram sizes of the different BV scales (Methods 3.4). The scale medians L (Equations (27) and (28)) have been chosen as the interval width of the number of image scales that are included in this interval. While $k(i)$ is intended to determine the degree of overlap between the bins, it is expressed through the initiation of each propagated interval; see Equation (27). The information of the different binary images of the scale information is, therefore, expected to be reorganized into better new partitions, which will follow the flexible size of the BV along their different locations.

$$\widehat{wb_k}(x,y) = Med(wb_{cc}\ (x,y,k(i)),\quad wb_{cc}\ (x,y,k(i)+1),\ldots wb_{cc}\ (x,y,k(i)+L)) \tag{26}$$

where L is the width of the median interval, and $i = 1 : N - L$.

The median is calculated as follows:

$$Med\ (X) = \begin{cases} X\left[\frac{L}{2}\right]\ if\ L\ is\ even \\ \frac{(X\left[\frac{L-1}{2}\right]+X\left[\frac{L+1}{2}\right])}{2}\ if\ L\ is\ odd \end{cases} \tag{27}$$

X is defined as the ordered list of values in the dataset, and L is the number of values in the dataset.

After the model's stage that suggested how to reorganize the different image scales, through the new intervals, $WB(x,y)$, Equation (29) shows how the different scale binary images $\widehat{wb_k}(x,y)$ are summed into a single multiscale ROI image (Figure 4C2).

$$WB(x,y) = \sum_{k=1}^{N} \widehat{wb_k}(x,y) \tag{28}$$

where N is the number of $\widehat{wb_k}(x,y)$ binary images.

Since $WB(x,y)$, Equation (28), reflects the summation of the values of the different scales from each location, this summation can obtain a response, which is beyond the value one at specific locations. This is derived from obtaining agreement of the same value from more than one scale. The significance of such results reflects a possibility that some BVs, at specific locations, are shared by more than a single scale. However, the response of $WB(x,y)$ should yield a binary image and, therefore, should yield only a response of zero or one. In order to overcome this issue, we chose to use a criterion that yields a response that acknowledges what will be the value at a specific location under the below conditions; see Equation (29). The criterion can be taken as the summed value τ (Methods), which reflects a chosen $\widehat{WB}(x,y)$ that will determine if $\widehat{WB}(x,y)$ will be finally determined as one or zero.

$$\widehat{WB}(x,y) = \begin{cases} 1\ if\ WB(x,y) > \tau \\ 0\ otherwise \end{cases} \tag{29}$$

where τ is the intensity threshold that will end up with final values for $\widehat{WB}(x,y)$, which will determine the ROI image $\widehat{WB}(x,y)$.

3.4. BV Texture

The output of the enhanced edges, T_{wb}, (the line completion responses) in the image T_E (Figure 3C, Equation (23)) can be referred, in this stage of the algorithm, only to the labeled ROI (Figure 4C2, Equation (30)). Therefore, $\widehat{WB}(x,y)$ (Equation (30) and Figure 5) is calculated to delineate the spatial location in which the line completion T_E is applied.

$$T_{wb} = \widehat{WB}(x,y) \times T_E \tag{30}$$

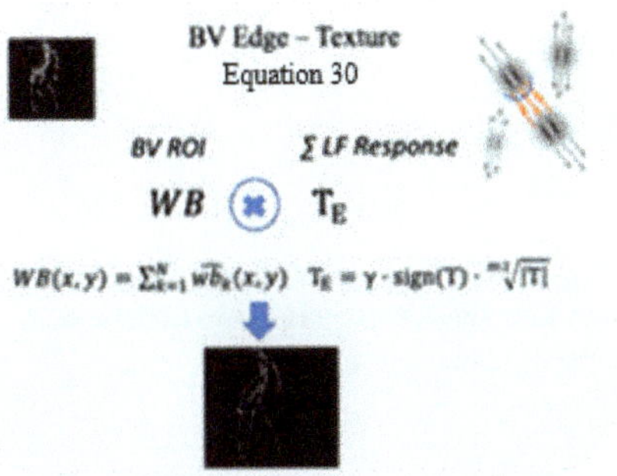

Figure 5. The flow chart illustrates the multiplication of the extracted ROI (**left** binary image) with the multiscale *LF* response. The right illustration demonstrates receptive fields with their summated additive signal (orange arrows). The figures represent the extracted ROI (**left** upper image) and the resulting blood vessel multiscale. edges (**bottom** image).

3.5. BV Visibility Enhancement

The BV edge image, T_{wb}, contains the enhanced edges within the relevant ROI's locations, but this image is not contained in the diverse DC details, which are not at the edges, such as in the input image $I(x,y)$. Before performing the integration of the edge image T_{wb} with the original image, we need to deal with the separation of the positive T_P and negative T_N values of the edge image T_{wb} (Equation (32)). The integration operation is completed also to enable an appropriate correction of the dynamic range of the final image $I_E(x,y)$ (Equation (33); [27]).

The separation of the blood vessels' edge image T_{wb} into positive T_P and negative T_N responses is shown below:

$$T_P = \max(0, T_{wb}),\ T_N = \max(0, -T_{wb}) \tag{31}$$

where T_P and T_N represents the absolute values of the positive and negative responses through the maximum operation.

The following stage, consequently, is the integration of the texture image of both the positive, T_P, and negative, T_N, blood vessel responses (Equation (32); Figure 6), with the input image $I(x,y)$ (Equation (33)), Figure 6. The weight function of each of the components, the original image, and the edges, can be determined by factor α. In order to improve the dynamic range image, we need to have a similar range of values for the positive and the negative components of the image, and this is achieved through the β factor, as shown in Equation (33). This β factor is intended to enable the balance of the bright and dark zones, for the $I_E(x, y)$ response.

$$I_E(x, y) = I(x,y) + \alpha \times \left[\frac{T_P}{1 + \beta \cdot I(x,y)} - \frac{T_N}{1 + \beta - \beta \cdot I(x,y)}\right] \quad (32)$$

where each polarity component is balanced by factor β. α, β are positive constants. The aim of the division factors is to suppress the positive or negative responses that are near the zero clipping, respectively.

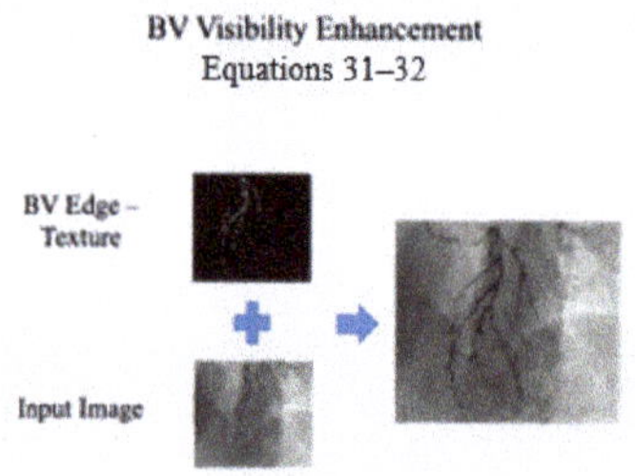

Figure 6. Schematic diagram of the last stage of the algorithm that integrates the edge image (**upper left** image) and the input image (**bottom left** image) into an enhanced visibility image (**right** image). An additional stage is performed, which is related to the dynamic range.

4. Methods

4.1. Dataset

Our clinical dataset consists of coronary cine-angiograms acquired during 3 different routine interventional catheterization procedures performed on 3 different patients in a single medical center. Altogether, 62 different runs were acquired, with an average of 40 frames per run. The overall number of coronary angiogram frames in the dataset is 2500, and the resolution of each image is 1024 × 1024 pixels, with 1024 grey levels per pixel. Our implementation and testing were performed in a MATLAB environment, using an *Intel® Core™* i9 2.3 GHz CPU laptop with 16 GB of main memory. The average processing time is 1.9 ± 0.05 s per frame.

4.2. Model Parameters

4.2.1. Orientation and Scale Filtering (Gabor)

The Gabor filter (Equation (1)) parameters are $\sigma = 8$, $\lambda = 12$.

Instead of using multiple Gabor filter sizes, the algorithm uses a fixed Gabor filter size, while deconstructing the input image into multiple scales, as is commonly used in Gaussian Image Pyramid deconstruction. The chosen number of scales is 8, and the number of Gabor filter orientations is 8. The chosen orientation angles, θ, are as follows: 0°–360° (Equation (1)).

Downscaling and upscaling of the images were completed by using the Lanczos resampling method. This method approximates the theoretically optimal reconstruction filtering for band-limited signals, which uses the Sinc filter. We used it for its anti-aliasing and preservation of sharp edge properties, which are important for maintaining the quality of the images, along the different algorithm stages.

The parameters used for the Naka–Rushton function (Equation (8)) is $\sigma_{NR}^{n} = 0.1$, $n = 2$.

The parameter used for the orientation recombination of the different images is m1 = 1 (Equations (20) and (21)). The parameters used for the scale recombination of different images are m2 = 2 and $\gamma = 1/8$ (Equations (22) and (23)).

4.2.2. Line Completion Algorithm

The values of σ_x, σ_y (Equations (10)–(12)) represent the magnitude of the additive signal. Equation (10)'s decay in the collinear axis and its orthogonal weight function of the signal, Equation (11), are defined separately for each scale (Table 1). The percentile used for the threshold of the line completion, γ_j (Equation (18)), is also detailed in Table 1.

Table 1. Algorithm parameters per scale.

Resolution Index (j)	σ_{xj}	σ_{y_j}	γ_j [%]	r_j $(pixels)$
1	0.05	0.05	99	0
2	0.05	0.05	80	2
3	0.06	0.06	97	3
4	0.07	0.07	97	4
5	0.07	0.07	96	5
6	0.07	0.07	96	5
7	0.1	0.1	96	5
8	0.1	0.1	96	6

Part of the algorithm's parameters has a different set of values per scale (j). These parameters are presented in Table 1. The background color has been added for identification the used resolutions for small vessels inside the cardinal vessels.

4.2.3. BV ROI

The structuring element disk radius, r_j $(pixels)$, is used for the closing procedure (Equations (25) and (26)) as detailed in Table 1. The width of the median filter is $L = 3$; see Equations (27) and (28). The resulting number of median-filtered scales, $\widehat{wb_k}(x, y)$, is 6.

4.2.4. BV Texture

The parameters used for integrating the enhanced structure signal and the original image are $\alpha = 0.4$ and $\beta = 4$ (Equation (32)).

4.3. Cardiologists' Evaluation of the Algorithm Results

All images were assessed by the consensus of 2 experienced interventional cardiologists (authors of our paper). First, viewing Diacom format cine runs of clinical coronary angiograms and choosing from each run (15 frames/s), we acquired, from standard routine angles, a single frame that best demonstrates the most significant clinical procedural details. These included general visibility of the whole frame, the CBVs, especially the small ones, the lesions within the BVs, and the various angiography interventional tools. Next, each cine run and each single chosen frame ("raw" image), as above, were visually compared, first with the same image after further manual correction for optimal visibility (modifying brightness and contrast) and, finally, with the improved results and the post-processing results of the novel algorithm.

5. Results

The algorithm results are presented according to the clinical implications. The figures in the current paper represent a partial sample from the overall results, which were selected by the cardiologists, mainly due to their diagnostic significance.

5.1. Overall Visibility Enhancement

Figure 7 demonstrates the overall improvement in the visibility of the BVs in the images derived from two different CAGs, from two patients. The result images show that the overall visibility of the BVs is enhanced with enhanced appearance. It can be seen at a naïve glance that the algorithm improved the dynamic range of the images, and therefore, the illumination and brightness of the images are more suitable for observing the structure and the richness of the vascular tree. The cardiologists who assessed the algorithm results considered the general visibility of all the CBVs as improved, with special emphasis on the better visibility of the small BVs, the collaterals, and the overlapping side branches (Figure 8).

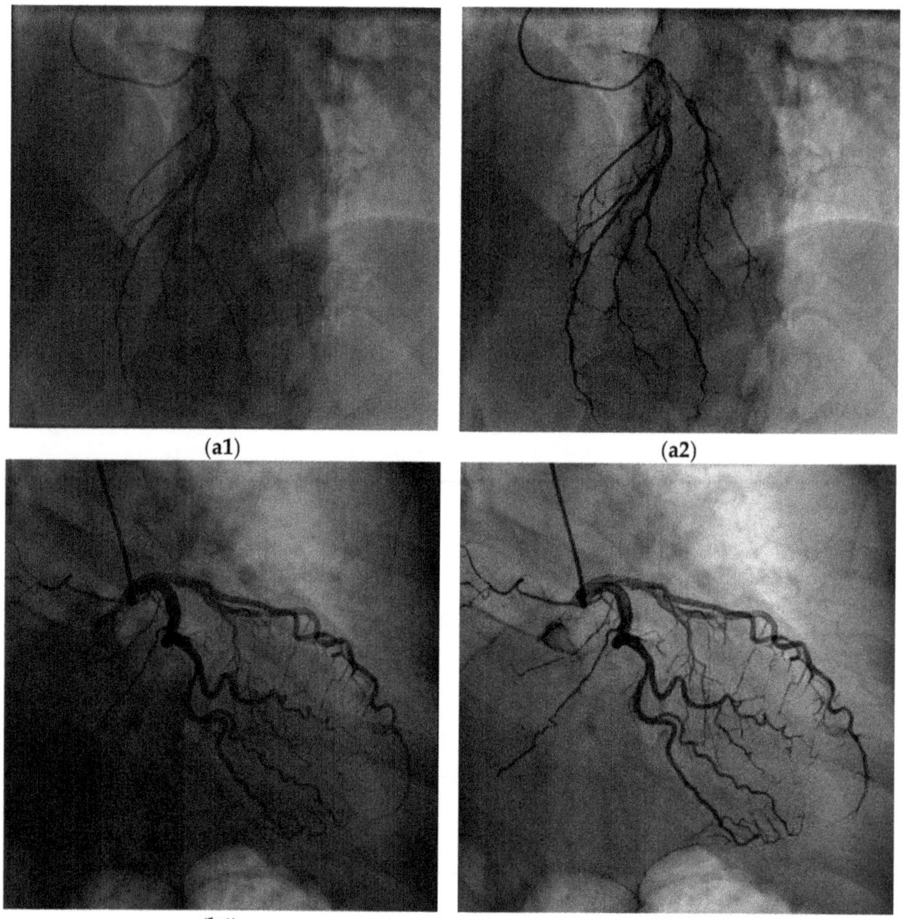

(a1) (a2) (b1) (b2)

Figure 7. Two examples of VIAEVCA results (**a2**,**b2**), which represent the better overall appearance of an angiogram with a better view of the BVs, including of the small ones. Original ("raw") images are shown on the left column and the algorithm results on the right column. This mode of presentation of the image results refers to all the figures in this study. (**a1**,**b1**) are the original images before the algorithm's applications.

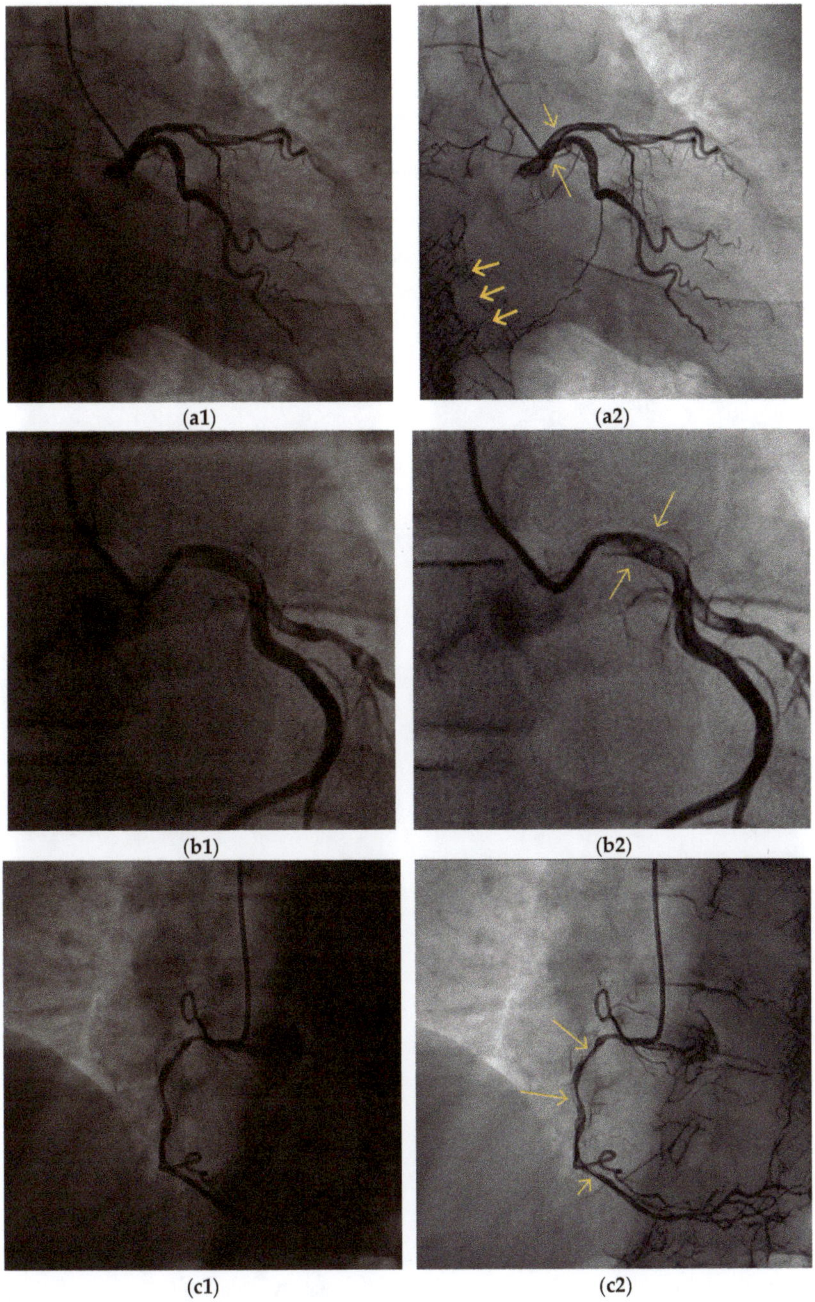

Figure 8. The VIAEVCA succeeds in increasing the visibility of the interior content of the BVs while revealing important clinical details. (a2) The upper long arrow indicates the exposure of the catheter tip, while the upper short arrow indicates exposure of the inner lumen of the left main coronary artery. The spine artifacts are exposed and are indicated by the thick left arrows. (b2) The BV shows overlapping vessels and lesions (see text); see the long arrows. (c2) Two dissections of the right coronary artery (long arrows) are exposed. The wire is also more visible (short arrow) with the results image. (a1,b1,c1) are the original images before the algorithm's applications.

5.2. Blood Vessel Interiors

One of the goals of our algorithm was to enable observing the BV's interior content and expose important clinical details, such as the catheter tip, catheterization apparatus, and inflections. Figure 8 (right column) demonstrates the exposure of the interior part of the major CBV, as represented by the long arrows. The cardiologists assessed the better visibility of the content within the lumen (the interior of the vessel) of the large BV, which enabled somewhat better identification of overlapping vessels (versus similar-looking luminal lesions) (Figure 8b, long arrows), coronary artery dissections (Figure 8c, long arrows), various interventional tools such as the tip of a diagnostic catheter (Figure 8a, long arrow), wires (Figure 8c, short arrow; Figure 9c, left arrow), balloons (Figure 10b,c long arrow), and stents (Figure 9a, long arrows; Figure 10a, long arrows).

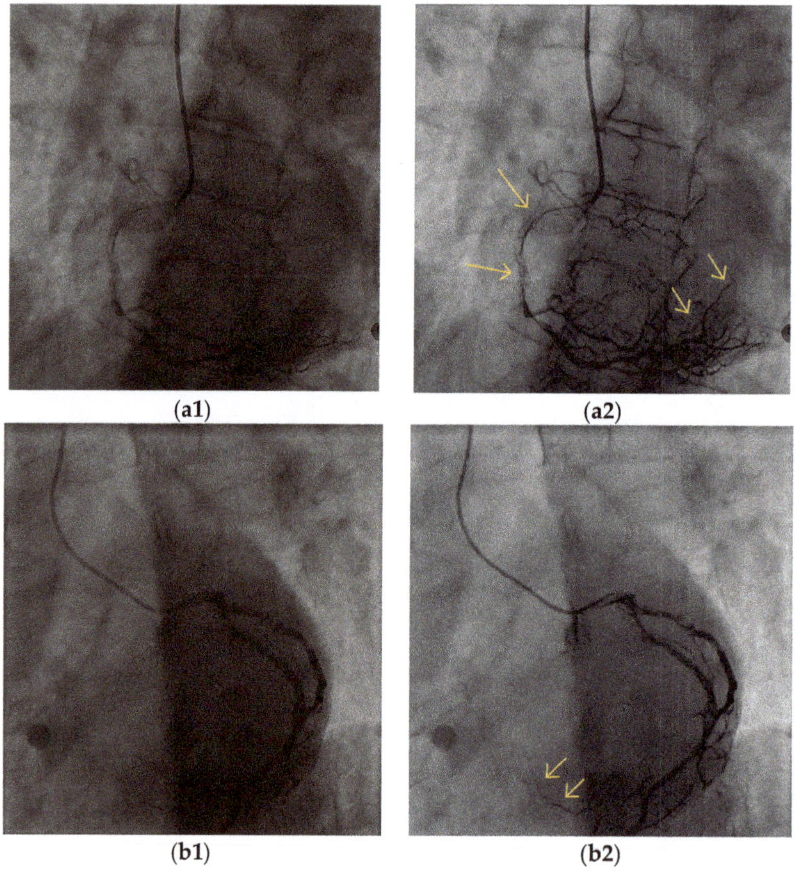

Figure 9. The VIAEVCA enables the better appearance of the small and clinically important blood vessels. (**a2**) The right arrows show better visibility of the small BVs, which split and climb upwards. Their diagnostic significance is described in the text. The left arrows show the two stents. (**b2**) The arrows show the exposed right coronary artery (see text for clinical significance). (**a1**,**b1**) are the original images before the algorithm's applications.

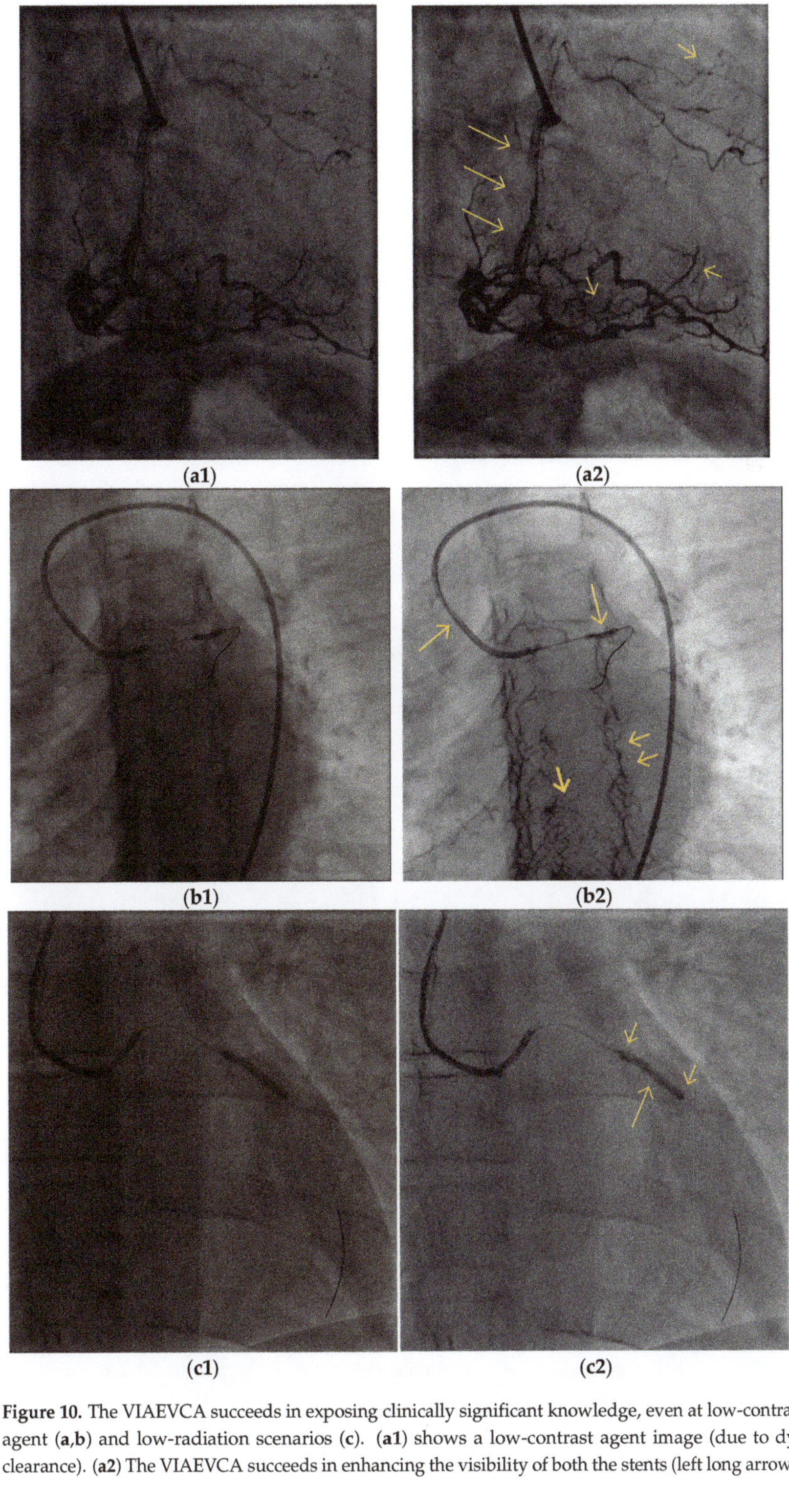

(a1) **(a2)**

(b1) **(b2)**

(c1) **(c2)**

Figure 10. The VIAEVCA succeeds in exposing clinically significant knowledge, even at low-contrast agent (**a**,**b**) and low-radiation scenarios (**c**). (**a1**) shows a low-contrast agent image (due to dye clearance). (**a2**) The VIAEVCA succeeds in enhancing the visibility of both the stents (left long arrows) and the small blood vessels (right short arrows). (**b1**) shows an image before contrast agent injection. (**b2**) The VIAEVCA succeeds in showing the wire inside the diagnostic catheter (left long arrow) and inside the CBV with the inflated balloon (right long arrow). Bony artifacts (thick and short arrows) are exposed in the region of the spine along with accentuation of edges of spinal bone structures. (**c1**) shows a low-radiation image (1/13 reduced radiation). (**c2**) The stent with its proximal and distal metal markings (short arrows); the wire that passes through it (long arrow) is better exposed.

5.3. Small Blood Vessels

A successful diagnosis by cardiologists requires proper observation of most of the blood vessels, including the small ones (Introduction). Figure 9 and additional figures, such as Figures 7 and 10a, demonstrate the exposure of a single or a net of small BVs with deficient exposure in the original image. Figure 9 demonstrates the ability of the algorithm to enhance the visibility of small BVs, which is of special clinical importance. The better exposure of a net of small BVs that feeds the interventricular septum (Figure 9a2, right arrows) can indicate a successful intervention in the right coronary artery stenting and replenishment of blood supply (Figure 9b2, left arrows). An enhanced small BV can be seen in Figure 9b2, where a collateral from the left to the right coronary artery is demonstrated. This helps to diagnose, from left system imaging, a proximal obstruction of the right CBV, since this collateral is not usually seen where the right CBV has good blood supply.

5.4. Low-Contrast Agent, Low-Radiation, and No Contrast Agent

We also tested angiography images with low radiation and low contrast, with the same algorithm that has the same set of parameters, which was used for the whole set of CAG images.

Our dataset lacks true low-contrast angiography procedures. Yet, we identified low-dose conditions where the injected contrast agent barely arrives at its specific locations or where the contrast agent starts to wash out and is partially cleared from the BVs. Thus, for this purpose, we used timed CAG images where the contrast agent is diluted in the BVs.

Figure 10 represents several examples from our dataset that can be regarded as acquired at low radiation (Figure 10c), at low contrast (due to dye clearance, Figure 10a), and in the absence of a contrast agent. The last group of images was acquired before the contrast agent injection (Figure 10b). Figure 10a1 presents an example of an image acquired during the clearance of the dye from the proximal part of a large CBV. VIAEVCAs enabled improved visibility of the major CBV content and the stents (Figure 10a2, long arrows). The algorithm also succeeded in enhancing the visibility of the small CBVs, although the dye started to be cleared distally at this stage.

Figure 10b refers to the CAG condition where the contrast agent was not yet injected into the BVs, and therefore, only the catheterization tools are exposed. The algorithm better exposed the wire in the internal side of the diagnostic catheter (left arrow) and inside the CBV and the attached balloon (perpendicular arrow). In this image, the artifacts of the bony structure are also exposed (thick and short arrows.)

The reduction in radiation to 1/13 per image (attempting to reduce tissue damage for the documentation of stent deployment) did not impair the advantage of the algorithm to enhance visibility (Figure 10c2), compared to the natural camera image (Figure 10c1) or even manually processing the same image in an optimal manner. Note also that the algorithm partially performs a compounding compression and expands the dynamic range of the image. This is expressed by the exposure of the image information in the too-dark and too-bright regions (this can be seen across all images).

Notably, although the algorithm succeeded in improving many aspects of the CAG images, it also enhances the appearance of disturbing vertebral and bony artifacts, which can appear as small blood vessels and may lower diagnostic accuracy, especially while viewing only still images (Figures 8a2 and 9a2). However, viewing the images from other camera angles and in the cine mode, which allows the evaluation of flow within the CBVs, mostly helped to clinically differentiate between those artifacts and true small vessels (Figure 10a2).

6. Discussion

Our novel algorithm has been inspired by the visual system mechanism for contour integration and lateral facilitation [27,29]. It succeeds in overcoming several critical CAG appearance and visibility challenges, such as small BVs, catheterization tools, and visibility under low-radiation and/or low-contrast agent conditions. The enhanced CAG visibility (Figures 7–10) led to improved images for better identification and classification of obstructive lesions, collateral blood vessels, the evaluation of flow, tissue perfusion, and catheterization tools. The algorithm also revealed an exposure of unwanted details of bony structures, which may interfere with the diagnosis of the small BVs. This drawback is expressed mainly while viewing still images acquired from limited angles but much less in the cine mode and at multiple angles.

The enhanced CAG visibility suggests a better decision-making process by the expert cardiologist before, during, and after the procedure. These achievements have been obtained with a single compound algorithm, which also succeeds in significantly enhancing images even with a low-dose contrast agent or low radiation, while simultaneously reducing noise and illumination problems. The advantages of the algorithm's performance are compared to the general and topic results of previous studies.

Small blood vessels: Several recent studies related directly to the issue of revealing small BVs [2,8,14]. Some previous studies related to the issue of small BVs as only a side effect of generally enhancing the image, through different methods, such as de-noising [9,15,19]. The noise reduction succeeded in enhancing the appearance of BVs, but it appears that the exposure of small BVs was not salient [9,15,19].

Two of the more recent segmentation studies applied a deep learning U-net [14] and decision tree [8] and succeeded in obtaining better results in reference to small BVs. This was achieved by cardiologists manually delineating this type of BV for the training dataset. Still, the visual impression of the small BV segmentation was considered insufficient [8,14].

It appears that the VIAEVCA exposure succeeded in revealing a larger repertoire of small BVs (Figures 7, 9a2, and 10a), along with bony artifacts, which, however, can be partially overcome, as described above. These artifacts are indeed prevented when a routine segmentation operation is performed. Nevertheless, our algorithm's enhanced visibility provides a much better ability to observe the small BVs, an important diagnostic advantage. The common goal of a successful invasive procedure is usually the restoration of flow in the obstructed CBV, especially under emergency circumstances such as acute myocardial infarction (AMI) or acute coronary syndrome (ACS). Yet, sometimes, despite epicardial patency after opening the artery, the restoration of good flow and myocardial perfusion is absent. This no-flow phenomenon has significant clinical importance for treatment and prognosis, but the current methods to clinically assess it (measuring or scoring flow and "myocardial blush") are limited, time-consuming, or expensive.

Future research may show a correlation between this no-flow phenomenon and the disappearance of small vessels, which were visible before the invasive intervention only by using the algorithm. Also, cardiac diseases such as syndrome X or metabolic cardiac syndrome, which are defined by typical anginal pain with myocardial ischemia but normal patent coronary arteries, may eventually show abnormal small vessel disease, detectible due to the enhanced visibility offered by the current novel algorithm. We could not find other algorithms that relate to these pathological issues.

Since previous algorithms [24,25] produced segmentation of the BVs, they could provide a quantitative measure for accuracy performance (in comparison to the manual cardiologist segmentation). One study [24] arrived at a 95% accuracy performance, based on seven images, which were examined. Ref. [14] used a larger dataset consisting of 48 images and arrived at a DICE performance of 76%. Our structure enhancement algorithm results cannot be compared quantitatively to the segmentation results since our algorithm does not produce BV segmentation. In fact, even the high DICE or accuracy performance, as above, does not necessarily reflect the results related to small BVs, since the area that they occupy in the image is relatively small. Additionally, the above studies [24,25] based their results on single images instead of cine images, which are commonly used in diagnostic CAGs. Our algorithm enabled clinical assessments based on consecutive images and the cine mode of the images, which can also reduce the effect of distractions caused by the bony artifacts.

Several groups presented algorithm results using the CNR measure. Lee and his colleagues [17] showed an 81% CNR improvement over the low-dose images (acquired using 2% of the radiation used in normal CAGs). They arrived at a higher CNR measure than our results, a 60.71% CNR improvement. They achieved a higher CNR score probably due to the noise reduction that better smoothed the background image but did not necessarily expose a larger number of small BVs (their images do not show a trend of a higher number of small BVs).

Studies that performed segmentation by background subtraction [18,20–22] yielded much better CNR results (~200% CNR improvement over the original) than studies that performed enhancement of visibility, as ours. This was achieved by the isolation of the BV image from the background and the obtained separation from other organs, such as the spinal cord, and probably, due to the fact that different organ components have different spatial-temporal subtraction properties [18,20–22]. The price for such sophisticated algorithms was the algorithms' complexity and performance durations. In any case, it is expected that the segmentation operation will lead to a criterion with a much higher CNR measure, which may not necessarily reflect all clinical demands and cannot really be compared to our visibility measure.

We further ask whether the measure of CNR, which reflects different average intensity values in a specific region in the BV in comparison to the background, is, indeed, an efficient measure for better visibility of the BV. It could be a type of legitimate measure if the only need would be to enhance the visibility of the major BV, but by increasing the visibility, we would like to achieve an additional two measure goals. One of them is to enhance the visibility of small BVs, for increasing diagnostic abilities (Section 4.3). This goal was reported by just one of the background reduction studies (Thuy et al., 2021 [25]). The second goal is to increase the visibility of the BV content, such as the CAG tools and plaques, enabling improved procedures. This BV exposure was probably derived from increasing the visibility of the texture inside the BV region (i.e., increasing the variance values) and not simply increasing the intensity values, as utilized by segmentation operations. Therefore, a common CNR improvement cannot be reflected by such a required property of the background subtraction results and their CNR-obtained measures. Nevertheless, future studies may combine these methods to enhance BVs in addition to improving visibility inside vessels using various CAG tools.

It has been reported by one background reduction study that, despite these algorithms' impressive results, the exposure of small BVs is still lacking [20], and also, the visibility of the catheterization tools inside the BVs is missing. Furthermore, these methods can lead to a condition where the small BVs can be fragmented into different layers (background and foreground, for example), due to matrix decomposition [20]. These authors suggested the application of Gabor wavelets to enable the enhancement of the tiny blood vessels [30] but did not test this possibility. Our algorithm can be used as a type of preprocessing for image enhancement and might be more beneficial than a Gabor wavelet (ref), due to its ability to perform line completion while decreasing the noise level.

Low Dose: A comparison of our algorithm results on low-dose conditions (low-radiation and low-contrast agents) is not sufficiently accessible, since there is only one paper [17] that provided algorithm results regarding low radiation at the CAG images. An additional two papers, which improved the visibility of the BVs in conditions mimicking low-dose radiation [1] and a low-contrast agent [18], showed an improved image with somewhat better visibility of the BVs.

To evaluate our algorithm results on the low-radiation images, we used images taken from a fluoroscopy run (used for the documentation of stent deployment), which routinely utilize only 1/13 of the radiation dose, relative to CAG images (Figure 10b2). The expert cardiologists assessed the algorithm results, which succeeded in enhancing the visibility of the procedural tools, such as the catheter tip, the wire, the balloon, and the stent edge markings (Figures 8–10). None of the previous studies showed such an ability to obtain visibility of the inner BV tools, neither under normal radiation and contrast agent doses nor at low-radiation or low-contrast agent conditions. We, therefore, propose our algorithm, which shows a potential usage of a CAG based on a low-radiation or low-contrast medium, which is less medically harmful. Future clinical studies using the algorithm are expected to validate the method and enable longer procedures, as needed, especially for more patients susceptible to damage, such as patients with renal dysfunction.

Inner blood vessel visibility: In the literature, there is a reference to catheter visibility and detection, but only outside the BVs [11], with no reference related to enhanced visibility of the catheterization tools in general. More specifically, previous algorithms that performed different algorithms for contrast enhancement, such as by fractional differential kernels [15] or Hessian matrices [19], did not refer to those tools.

We also tested whether an algorithm that has the potential to expose the dynamic range of specific areas in the image, such as the CLAHE algorithm (Matlab), has the potential to expose the catheter tools in the CBVs. We found some BV enhancement, but with significantly reduced visibility of the tools than our algorithm presents.

The VIAEVCA algorithm is probably the first to enable distinct enhanced visibility of the CAG procedural tools inside the major BVs. This achievement was obtained through an innovative approach to object enhancement utilizing the lateral facilitation mechanism component and additional computations regarding BV texture. Since our study, in its present form, does not perform segmentation or subtraction operations, it cannot compete with published background subtraction results and their CNR-obtained measures. Nevertheless, future studies may combine these methods to enhance BVs in addition to improving visibility inside the vessels using various CAG tools.

7. Conclusions

The present study presents a novel algorithm for enhancing BV visibility from routine angiography video images.

Original algorithm: (a) This algorithm was inspired by an educated guess mechanism suggested for the visual phenomenon of lateral facilitation. It causes line and texture completion imitating our visual perception and is expressed by the lateral facilitation phenomenon. (b) The novelty of the algorithm is also expressed by incorporating partial noise reduction simultaneously with line completion, as part of strengthening the structure.

Algorithm achievements: (a) This algorithm enhanced the visibility of the BVs, including the small BVs. (b) This is the first algorithm that demonstrates enhanced visibility of major BV luminal content, including CAG procedural tools and plaque lesions. (c) Through the algorithm adaptivity, it also succeeds in enhancing BV signals at low doses of contrast agents and radiation, with the same set of parameters. These achievements contribute together to support the clinician in making an accurate diagnosis, better procedural performance, and improved patient outcomes.

Author Contributions: Conceptualization, H.S. and J.E.; Methodology, H.S. and J.E.; Software, Y.S.K. and H.S.; Validation, J.E., M.M. and H.S.; Formal analysis, H.S. and Y.S.K.; Investigation, H.S.; Re-sources, Y.S.K., H.S., J.E. and M.M.; Data curation, J.E. and M.M.; Writing, H.S., Y.S.K. and J.E.; Writing-review and editing, H.S. and J.E.; Visualization, H.S.; Supervision, H.S., J.E. and M.M.; Project administration, H.S.; funding acquisition, H.S. All authors have read and agreed to the published version of the manuscript.

Funding: This research received no external funding.

Institutional Review Board Statement: Not applicable.

Data Availability Statement: The original contributions presented in this study are included in the article. Further inquiries can be directed to the corresponding author.

Conflicts of Interest: The authors declare no conflict of interest.

References

1. Pavoni, M.; Chang, Y.; Park, S.; Smedby, O. Convolutional neural network-based image enhancement for X-Ray percutaneous coronary intervention. *J. Med. Imaging* **2018**, *5*, 1. [CrossRef] [PubMed]
2. Kato, S.; Kitagawa, K.; Ishida, N.; Ishida, M.; Nagata, M.; Ichikawa, Y.; Katahira, K.; Matsumoto, Y.; Seo, K.; Ochiai, R.; et al. Assessment of coronary artery disease using magnetic resonance coronary angiography: A national multicenter trial. *J. Am. Coll. Cardiol.* **2010**, *56*, 983–991. [CrossRef]
3. Hwang, J.; Lee, S.Y.; Chon, M.K.; Lee, S.H.; Hwang, K.W.; Kim, J.S.; Park, Y.H.; Kim, J.H.; Chun, K.J. Radiation exposure in coronary angiography: A comparison of cineangiography and fluorography. *Korean Circ. J.* **2015**, *45*, 451–456. [CrossRef] [PubMed]
4. Moccia, S.; De Momi, E.; El Hadji, S.; Mattos, L.S. Blood vessel segmentation algorithms—Review of methods, datasets and evaluation metrics. *Comput. Methods Programs Biomed.* **2018**, *158*, 71–91. [CrossRef] [PubMed]
5. Zhao, F.; Chen, Y.; Hou, Y.; He, X. Segmentation of blood vessels using rule-based and machine-learning-based methods: A review. *Multimed. Syst.* **2019**, *25*, 109–118. [CrossRef]
6. Frangi, A.F.; Niessen, W.J.; Vincken, K.L.; Viergever, M.A. Multiscale vessel enhancement filtering. (including Subser. Lect. Notes Artif. Intell. Lect. Notes Bioinformatics). *Lect. Notes Comput. Sci.* **1998**, *1496*, 130–137. [CrossRef]
7. Liu, Y.; Wan, W.; Zhang, X.; Liu, S.; Liu, Y.; Liu, H.; Zeng, X.; Wang, W.; Zhang, Q. Segmentation and Automatic Identification of Vasculature in Coronary Angiograms. *Comput. Math. Methods Med.* **2021**, *1*, 2747274. [CrossRef] [PubMed]
8. Samuel, P.M.; Veeramalai, T. VSSC Net: Vessel Specific Skip chain Convolutional Network for blood vessel segmentation. *Comput. Methods Programs Biomed.* **2021**, *198*, 105769. [CrossRef]
9. Wilkinson, M.; Westenberg, M.A. Shape Preserving Filament Enhancement Filtering Michael. In Proceedings of the 4th Medical Image Computing and Computer-Assisted Intervention Conference, Utrecht, The Netherlands, 14–17 October 2001; p. 2208. [CrossRef]
10. Carballal, A.; Novoa, F.J.; Fernandez-Lozano, C.; García-Guimaraes, M.; Aldama-López, G.; Calviño-Santos, R.; Vazquez-Rodriguez, J.M.; Pazos, A. Automatic multiscale vascular image segmentation algorithm for coronary angiography. *Biomed. Signal Process. Control* **2018**, *46*, 1–9. [CrossRef]
11. Fazlali, H.R.; Karimi, N.; Soroushmehr, S.M.R.; Shirani, S.; Nallamothu, B.K.; Ward, K.R.; Samavi, S.; Najarian, K. Vessel segmentation and catheter detection in X-ray angiograms using superpixels. *Med. Biol. Eng. Comput.* **2018**, *56*, 1515–1530. [CrossRef]

12. Felfelian, B.; Fazlali, H.R.; Karimi, N.; Soroushmehr, S.M.R.; Samavi, S.; Nallamothu, B.; Najarian, K. Vessel segmentation in low contrast X-Ray angiogram images. In Proceedings of the IEEE International Conference on Image Processing, Phoenix, AZ, USA, 25–28 September 2016.
13. Wan, T.; Shang, X.; Yang, W.; Chen, J.; Li, D.; Qin, Z. Automated coronary artery tree segmentation in X-ray angiography using improved Hessian based enhancement and statistical region merging. *Comput. Methods Programs Biomed.* **2018**, *157*, 179–190. [CrossRef] [PubMed]
14. Truc, P.T.H.; Khan, M.A.U.; Lee, Y.K.; Lee, S.; Kim, T.S. Vessel enhancement filter using directional filter bank. *Comput. Vis. Image Underst.* **2009**, *113*, 101–112. [CrossRef]
15. Pearl Mary, S.; Thanikaiselvan, V. Unified adaptive framework for contrast enhancement of blood vessels. *Int. J. Electr. Comput. Eng.* **2020**, *10*, 767–777. [CrossRef]
16. Pizer, S.M. Contrast-Limited Adaptive Histogram Equalization: Speed and Effectiveness. In Proceedings of the First Conference on Visualization in Biomedical Computing, Atlanta, GA, USA, 22–25 May 1990.
17. Lee, Y.G.; Lee, J.; Shin, Y.G.; Kang, H.C. Low-dose 2D X-ray angiography enhancement using 2-axis PCA for the preservation of blood-vessel region and noise minimization. *Comput. Methods Programs Biomed.* **2016**, *123*, 15–26. [CrossRef] [PubMed]
18. Ma, H.; Hao, H.; van Walsum, T. Vessel layer separation in x-ray angiograms with fully convolutional network. In Proceedings of the SPIE Medical Imaging: Image-Guided Procedures, Robotic Interventions, and Modeling, Huston, TX, USA, 10–15 February 2018; Volume 10576, pp. 443–457. [CrossRef]
19. Li, B.; Xie, W. Adaptive fractional differential approach and its application to medical image enhancement. *Comput. Electr. Eng.* **2015**, *45*, 324–335. [CrossRef]
20. Jin, M.; Li, R.; Jiang, J.; Qin, B. Extracting contrast-filled vessels in X-ray angiography by graduated RPCA with motion coherency constraint. *Pattern Recognit.* **2017**, *63*, 653–666. [CrossRef]
21. Song, S.; Du, C.; Ai, D.; Huang, Y.; Song, H.; Wang, Y.; Yang, J. Spatio-temporal constrained online layer separation for vascular enhancement in X-ray angiographic image sequence. *IEEE Trans. Circuits Syst. Video Technol.* **2020**, *30*, 3558–3570. [CrossRef]
22. Song, S.; Frangi, A.F.; Yang, J.; Ai, D.; Du, C.; Huang, Y.; Song, H.; Zhang, L.; Han, Y.; Wang, Y. Patch-Based Adaptive Background Subtraction for Vascular Enhancement in X-Ray Cineangiograms. *IEEE J. Biomed. Heal. Inform.* **2019**, *23*, 2563–2575. [CrossRef] [PubMed]
23. Cervantes-Sanchez, F.; Cruz-Aceves, I.; Hernandez-Aguirre, A.; Hernandez-González, M.A.; Solorio-Meza, S.E. Deep learning for coronary artery segmentation in x-ray angiograms using a patch-based training. In Proceedings of the 16th International Symposium on Medical Information Processing and Analysis, Lima, Peru, 3–4 October 2020. [CrossRef]
24. Rodrigues, E.O.; Rodrigues, L.O.; Lima, J.J. X-Ray cardiac angiographic vessel segmentation based on pixel classification using machine learning and region growing. *Biomed. Phys. Eng. Express* **2021**, *7*, 055026. [CrossRef]
25. Thuy, L.N.L.; Trinh, T.D.; Anh, L.H.; Kim, J.Y.; Hieu, H.T.; Bao, P.T. Coronary Vessel Segmentation by Coarse-to-Fine Strategy Using U-nets. *Biomed Res. Int.* **2021**, *1*, 5548517. [CrossRef] [PubMed]
26. Zhu, X.; Cheng, Z.; Wang, S.; Chen, X.; Lu, G. Coronary angiography image segmentation based on PSPNet. *Comput. Methods Programs Biomed.* **2021**, *200*, 105897. [CrossRef] [PubMed]
27. Matichin, H.; Einav, S.; Spitzer, H. A single additive mechanism predicts Lateral Interactions effects- computational model. *J. Opt. Soc. Am.* **2015**, *32*, 2247–2259. [CrossRef]
28. Polat, U.; Sagi, D. Lateral interactions between spatial channels: Suppression and facilitation revealed by lateral masking experiments. *Vis. Res.* **1993**, *33*, 993–999. [CrossRef] [PubMed]
29. Field, D.J.; Hayes, A.; Hess, R.F. Contour integration by the human visual system: Evidence for a local "association field". *Vis. Res.* **1993**, *33*, 173–193. [CrossRef] [PubMed]
30. Yin, B.; Li, H.; Sheng, B.; Hou, X.; Chen, Y.; Wu, W.; Li, P.; Shen, R.; Bao, Y.; Jia, W. Vessel extraction from non-fluorescein fundus images using orientation-aware detector. *Med. Image Anal.* **2015**, *26*, 232–242. [CrossRef]

biomimetics

MDPI

Article

Artificial Visual System for Stereo-Orientation Recognition Based on Hubel-Wiesel Model

Bin Li [1], Yuki Todo [2,*] and Zheng Tang [3,4,*]

[1] Division of Electrical Engineering and Computer Science, Kanazawa University, Kanazawa-shi 920-1192, Japan; crislee@stu.kanazawa-u.ac.jp

[2] Faculty of Electrical, Information and Communication Engineering, Kanazawa University, Kanazawa-shi 920-1192, Japan

[3] Institute of AI for Industries, Chinese Academy of Sciences, 168 Tianquan Road, Nanjing 211100, China

[4] School of Computer Engineering and Science, Shanghai University, Shanghai 200444, China

* Correspondence: yktodo@se.kanazawa-u.ac.jp (Y.T.); tang@iaii.ac.cn (Z.T.)

Abstract: Stereo-orientation selectivity is a fundamental neural mechanism in the brain that plays a crucial role in perception. However, due to the recognition process of high-dimensional spatial information commonly occurring in high-order cortex, we still know little about the mechanisms underlying stereo-orientation selectivity and lack a modeling strategy. A classical explanation for the mechanism of two-dimensional orientation selectivity within the primary visual cortex is based on the Hubel-Wiesel model, a cascading neural connection structure. The local-to-global information aggregation thought within the Hubel-Wiesel model not only contributed to neurophysiology but also inspired the development of computer vision fields. In this paper, we provide a clear and efficient conceptual understanding of stereo-orientation selectivity and propose a quantitative explanation for its generation based on the thought of local-to-global information aggregation within the Hubel-Wiesel model and develop an artificial visual system (AVS) for stereo-orientation recognition. Our approach involves modeling depth selective cells to receive depth information, simple stereo-orientation selective cells for combining distinct depth information inputs to generate various local stereo-orientation selectivity, and complex stereo-orientation selective cells responsible for integrating the same local information to generate global stereo-orientation selectivity. Simulation results demonstrate that our AVS is effective in stereo-orientation recognition and robust against spatial noise jitters. AVS achieved an overall over 90% accuracy on noise data in orientation recognition tasks, significantly outperforming deep models. In addition, the AVS contributes to enhancing deep models' performance, robustness, and stability in 3D object recognition tasks. Notably, AVS enhanced the TransNeXt model in improving its overall performance from 73.1% to 97.2% on the 3D-MNIST dataset and from 56.1% to 86.4% on the 3D-Fashion-MNIST dataset. Our explanation for the generation of stereo-orientation selectivity offers a reliable, explainable, and robust approach for extracting spatial features and provides a straightforward modeling method for neural computation research.

Keywords: stereo-orientation selectivity; Hubel-Wiesel model; artificial visual system

1. Introduction

Human visual perception of the external environment relies on various fundamental feature recognition processes, including motion, color, and orientation [1,2]. These primary visual feature extractions predominantly occur in the primary visual cortex and are subsequently integrated into the higher-level cortex for abstract feature recognition [3,4]. Among

 https://doi.org/10.3390/biomimetics10010038

these fundamental visual features, orientation feature response plays a crucial role in object edge detection, texture analysis, and shape recognition [5,6]. The neural mechanism of selective response to stimuli with preferred orientation is referred to as orientation selectivity [7]. Given this neuron property, many researchers conducted neural modeling based on orientation selectivity and inspired the development of artificial visual systems toward two-dimensional vision tasks [8–10]. As autonomous driving and robotics continue to advance, there is an increasing demand for processing three-dimensional spatial information inputs [11,12]. Therefore, investigating and modeling the neural mechanisms of spatial information recognition is beneficial for implementing robust information processing systems and advancing computational neuroscience.

The most remarkable study on orientation selectivity was conducted by Hubel and Wiesel, who observed several types of cortical neurons in the mammalian primary visual cortex exhibiting selective responses to visual stimuli [13,14]. They employed the terms V1 simple cells and complex cells to describe two typical neuron types with orientation selectivity [15,16]. Simple cells have relatively small receptive fields and show a preference for specific stimulus patterns [14]. These cells are vigorously activated only by stimuli with a particular orientation within their receptive field, while they exhibit no or weaker responses to other orientations [15]. On the other hand, complex cells also demonstrate selectivity towards stimulus orientation but possess larger receptive fields [15]. Unlike simple cells, complex cells maintain their response when preferred orientation stimuli move within their receptive field [14]. Hubel and Wiesel suggested that the generation of orientation selectivity originates from the hierarchical structure between LGN (lateral geniculate nucleus) and V1 (primary visual cortex) [17,18]. Specifically, V1 simple cells receive inputs from multiple LGN cells whose spatially close receptive fields are organized in a specific orientation pattern [3,19]. This arrangement leads to the development of corresponding orientation selectivity in simple cells. The complex cell is connected with those simple cells with spatially close receptive fields that share the same preferred orientation, resulting in a larger receptive field for complex cell responses while maintaining their selectivity to preferred stimuli orientations [17,20]. This explanation for the generation structure of orientation selectivity is commonly referred to as the Hubel-Wiesel model, which is extensively utilized in artificial visual system modeling [3,8–10,21]. The Hubel-Wiesel model is a concise and generalized concept that explains the feature extraction and aggregation process within the brain. Its thought of cascade connection and local-to-glocal information aggregation propelled the development of computational neuroscience [3,22–26] and computer vision [9,10,27–30].

The recognition of 3D information primarily occurs in the high-level visual cortical area [3,31,32]. Within these higher-order cortical areas, numerous neurons exhibit sensitivity to various 3D information, such as spatial depth, shape, curvature, surface characteristics, 3D fragments, and disparity [33–40]. While some researchers have reported that certain V4 neurons display stereo-orientation tuning without additional depth cues, it is more commonly observed that a substantial number of neurons exhibit cooperative responses to distinct spatial information [35,39,41]. The complexity associated with spatial information poses challenges in modeling spatial feature selectivity. However, it is worth noting that, similar to the generation process of two-dimensional orientation selectivity described in the Hubel-Wiesel model, the feature selectivity for spatial information in high-order cortical neurons also arises from the convergence of low-level simple features [8,19,42–46]. This cascading property offers a potential method for quantitatively reconstructing the neural mechanism of spatial feature selectivity [46–48].

Current research on stereo-orientation selectivity still focuses on discovering and observing the neuron activation or simulating the neuron activation and neural circuit connection [46,49–52]. There are limited attempts to develop a concise modeling scheme and further explore its application in stereo-orientation recognition. In addition, prior orientation recognition-related work mainly focuses on two-dimensional orientation, and detection techs include deep learning methods and traditional feature-matching methods [53–59]. Traditional techniques commonly have high computational complexity and poor adaptability to complex scenes [60]. Accordingly, the deep learning method is gradually being widely employed for object orientation detection tasks. The deep learning methods could obtain good performance after training with large amounts of data but require high computing resources and also lack robustness to adverse environmental conditions [30].

Enhancing the performance of systems in adverse environmental conditions is also a significant research topic [26,61,62]. Techniques for this issue include image preprocessing and enhancement [62–65], domain adaptation and transfer learning [66,67], adversarial training and self-supervised learning [68,69], and multimodal fusion [70,71]. In addition, biomimetic approaches also demonstrate significant potential in improving system performance [72,73]. For instance, brain cognitive computing can be employed to optimize the prediction of the flow state of the rectifier [74], using the Kolmogorov-Arnold network to predictive model electrohydrodynamic pumping (a biomimetic system) with interpretability [75], and biological visual mechanism can be applied to construct robust and explainable bio-inspired networks [76]. Especially in the image recognition task, introducing the biological visual mechanism into the construction of deep neural networks is also demonstrated as an efficient method to improve deep models' robustness and overall performance [30,76,77]. However, most bio-inspired neural networks concentrated on two-dimensional vision and image task, there is a lack of application and innovation in three-dimensional tasks. On one hand, spatial information differs from image information, and traditional Gabor filter-based methods are unsuitable for 3D tasks. On the other hand, we still know little about the recognition mechanism of spatial information within high-order visual cortical layers.

To extend the application of feature selectivity neuron properties to three-dimensional tasks, we propose a concise but efficient modeling strategy for stereo-orientation selectivity based on local-to-global information aggregation within the cascade Hubel-Wiesel model and develop an artificial visual system (AVS) for three-dimensional recognition tasks. We suggest initially employing depth selective cells to extract diverse depth information. Subsequently, the various depth information extracted by depth selective cells is combined in simple stereo-orientation selective cells to generate different local stereo-orientation selectivity. Lastly, the outputs from the simple stereo-orientation selective cells with identical stereo-orientation selectivity are integrated into a complex stereo-orientation selective cell, resulting in global stereo-orientation selectivity. Extensive simulation results demonstrate that our proposed mechanism of stereo-orientation selectivity is reliable, while the AVS based on this mechanism proves effective and stable in recognizing orientation information. Furthermore, compared to extensive deep neural network methods, our AVS exhibits considerable superiority in robustness against spatial noise jitters. The contributions of this paper are:

- We proposed a straightforward and concise modeling strategy for the generation of stereo-orientation selectivity and developed an AVS for stereo-orientation selectivity.
- The AVS is demonstrated with robustness, superiority, and efficiency in object oreintation detection tasks compared with deep models.
- Implementing AVS connected with deep models helps enhance deep models's performance, robustness, and stability in 3D object classification tasks.
- Our work provides a robust spatial feature pre-extraction method for 3D object recognition tasks and a potential explanation of how the brain perceives and encodes stereo-orientation information.

2. Methods

In this section, we introduce the mechanism of stereo-orientation selectivity and details of the artificial visual system (AVS) for stereo-orientation recognition. Firstly, we provide explanations regarding the generation of stereo-orientation selectivity based on the Hubel-Wiesel model. Subsequently, we give the modeling method of simple cells and complex cells. Lastly, we describe the implementation process of AVS.

2.1. Hubel-Wiesel Model and Stereo-Orientation Selectivity

The Hubel-Wiesel model describes the primary visual pathway as a cascade structure to explain the generation of orientation selectivity. Figure 1 illustrates the structure of the Hubel-Wiesel model. This model exhibits a hierarchical cascade architecture, wherein simple cells process local orientation information while complex cells integrate outputs from simple cells to encode more complex features. Specifically, the simple cell receives input from the LGN cells, whose receptive fields are arranged in a specific way that makes the simple cell sensitive to stimuli with specific orientations. Complex cell integrates inputs from multiple simple cells within an extensive receptive field to generate global orientation selectivity.

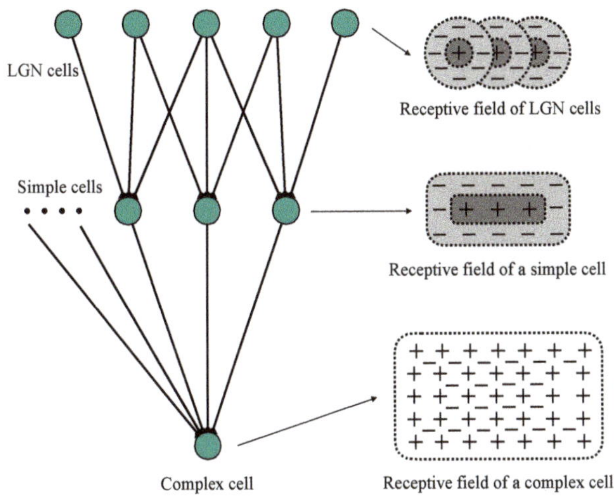

Figure 1. The hierarchical cascade structure of the Hubel-Wiesel model.

Previous studies have indicated that spatial features such as shape and curvature are derived from the integration of distinct depth information [78–80]. Drawing inspiration from the hierarchical cascade structure of the Hubel-Wiesel model, which emphasizes sequential processing and information integration, we propose an explanation for the generation of stereo-orientation selectivity. We suggest integrating multiple depth cues arranged in a specific spatial configuration to obtain stereo-orientation perception. Specifically, we initially employ depth selective cells to extract various positional information related to depth. Subsequently, each type of simple stereo-orientation selective cell receives inputs from several depth selective cells whose extracted depth information can be organized into a specific orientation within local space, thereby generating local stereo-orientation selectivity. Lastly, all simple stereo-orientation selective cells with the same orientation preference output to a complex stereo-orientation selective cell, resulting in the generation of global orientation selectivity. The conceptually neural connection underlying stereo-orientation selectivity is presented in Figure 2.

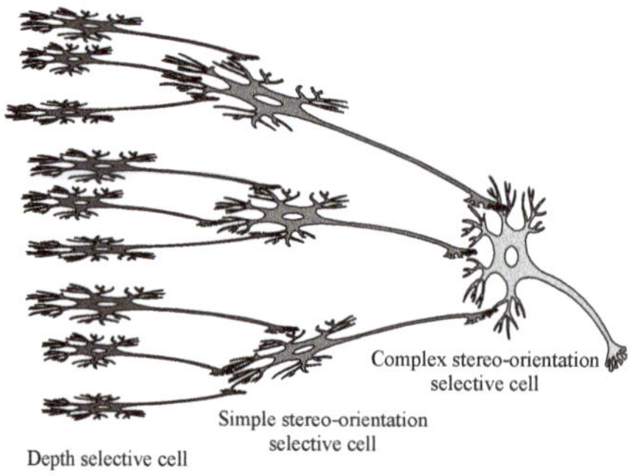

Figure 2. The conceptually neural connections of stereo-orientation selectivity.

2.2. Simple Stereo-Orientation Selective Cell

Local orientation selectivity is generated by simple stereo-orientation selective cells through the integration of specific depth information within a defined local space, which is extracted by depth selective cells. To quantify the computation of stereo-orientation selectivity, a fixed-scale local space with dimensions of $3 \times 3 \times 3$ is designed, allowing for the definition of 13 specific stereo-orientations within each local space. Figure 3 illustrates an instance of the information processing flow for local stereo-orientation information.

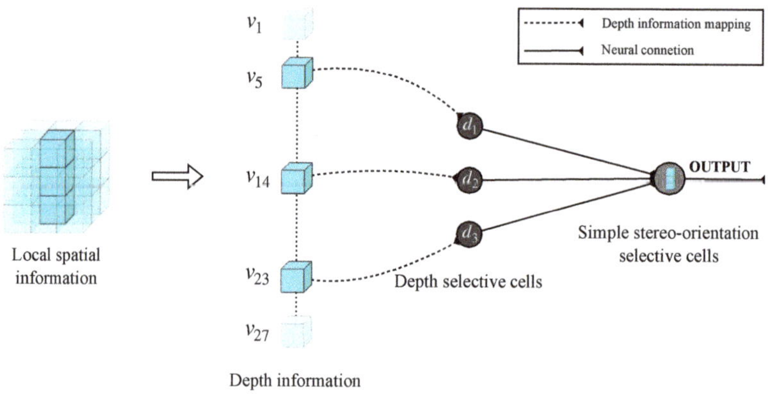

Figure 3. The information processing flow for local stereo-orientation information.

The depth selective cells do not directly accept the spatial information and are commonly reported as existing in high-order cortical layers [44,81–83]. There are multiple middle stages before depth selective cells respond to preferred depth information. However, the presence of depth information will truly activate specific depth selective cells, establishing a definitive causal relationship [3,82,83]. Therefore, the recognition process of spatial depth information can be directly simplified by mapping the voxel value of spatial elements at specific positions within a local space as direct inputs to depth selective cells. Within a local voxel space sized as $3 \times 3 \times 3$, 27 spatial elements can be identified and denoted from v_1 to v_{27}. Accordingly, the activation d of a depth selective cell can be quantitatively defined as follows:

$$d = v_{index}\, ,\, v_{index} = \begin{cases} 1 & \text{, if spatial element existed;} \\ 0 & \text{, otherwise.} \end{cases} \quad (1)$$

v_{index} denotes the voxel value (0 or 1) of a specific spatial element.

The simple stereo-orientation selective cells receive inputs from three specific depth selective cells, which provide corresponding depth information that can be combined to form a local stereo-orientation feature. As illustrated in Figure 3, the three depth selective cells extract distinct depth information, which collectively contributes to a vertical orientation feature. Subsequently, the simple stereo-orientation selective cell integrates these inputs and exhibits selectivity towards vertical orientation. We employed the Sigmoid function to model simple cell activation, a widely employed technique in the field of neural modeling [84–86]. The activation value s of a simple stereo-orientation selective cell is defined by the following formula:

$$s = \frac{1}{1 + e^{-k(\sum \mathbf{d_i} - \theta)}}, \tag{2}$$

where k denotes the response sensitivity of the cell, default as 10^4. $\mathbf{d_i}$ represents the inputs of each depth selective cell, and θ is the activation threshold, set to 2.5. We designed 13 distinct types of simple stereo-orientation selective cells, each corresponding to a specific local stereo-orientation defined within a local space. With the exception of the vertical orientation illustrated in Figure 3, the remaining 12 types of local stereo-orientation are depicted in Figure 4.

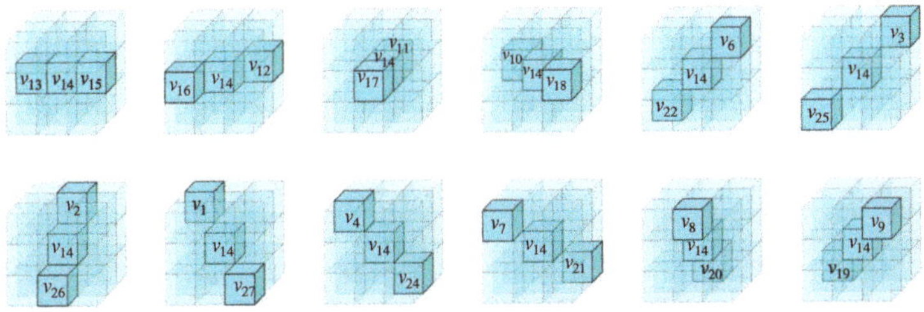

Figure 4. The remaining 12 types of stereo-orientation and their element indices within a local space.

2.3. Complex Stereo-Orientation Selective Cell

Based on our explanation for the mechanism of stereo-orientation selectivity, the complex stereo-orientation selective cells are responsible for the integration of local stereo information extracted by simple stereo-orientation selective cells. The connection pattern between simple and complex stereo-orientation selective cells is illustrated in Figure 5. We define the response rule of complex stereo-orientation selective cells as follows:

$$c = \sum_{i=1}^{N} s_i, \tag{3}$$

where c denotes the activation value of complex stereo-orientation selective cell, and N represents the total number of simple stereo-orientation selective cells. s_i denotes the inputs of each simple stereo-orientation selective cell.

2.4. Artificial Visual System for Stereo-Orientation Recognition

We consider that spatial information is organized in a similar way as image information. The global feature is the combination and connection of local feature fragments, and each fragment can be continually separated into multiple dependent spatial elements. Inversely, we can extract all local feature fragments by recognizing specific spatial elements and subsequently obtain the global features. Based on such thought of local-to-global information aggregation, we implemented an artificial visual system (AVS) for stereo-orientation recognition tasks. The overall process flow of AVS for object orientation recognition is depicted in Figure 6. For the object, we propose first taking each element of the object as the central point to segment out multiple local spaces. Subsequently, we employ a group of simple stereo-orientation selective cells consisting of 13 types that correspond to the defined local stereo-orientations within each local space to recognize the local stereo-orientation. Finally, these simple stereo-orientation selective cells are connected to complex stereo-orientation selective cells with identical stereo-orientation selectivity. The complex stereo-orientation selective cells integrate inputs from each cell group and serve as the output of AVS. The complex stereo-orientation selective cell with the greatest activation value indicates global object stereo-orientation.

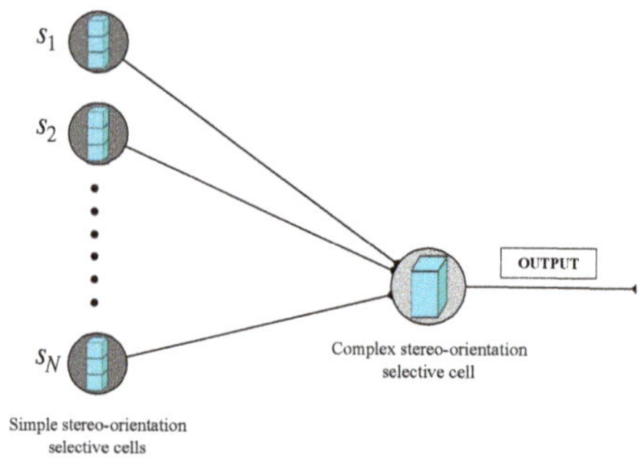

Figure 5. The connection pattern between simple and complex stereo-orientation selective cell.

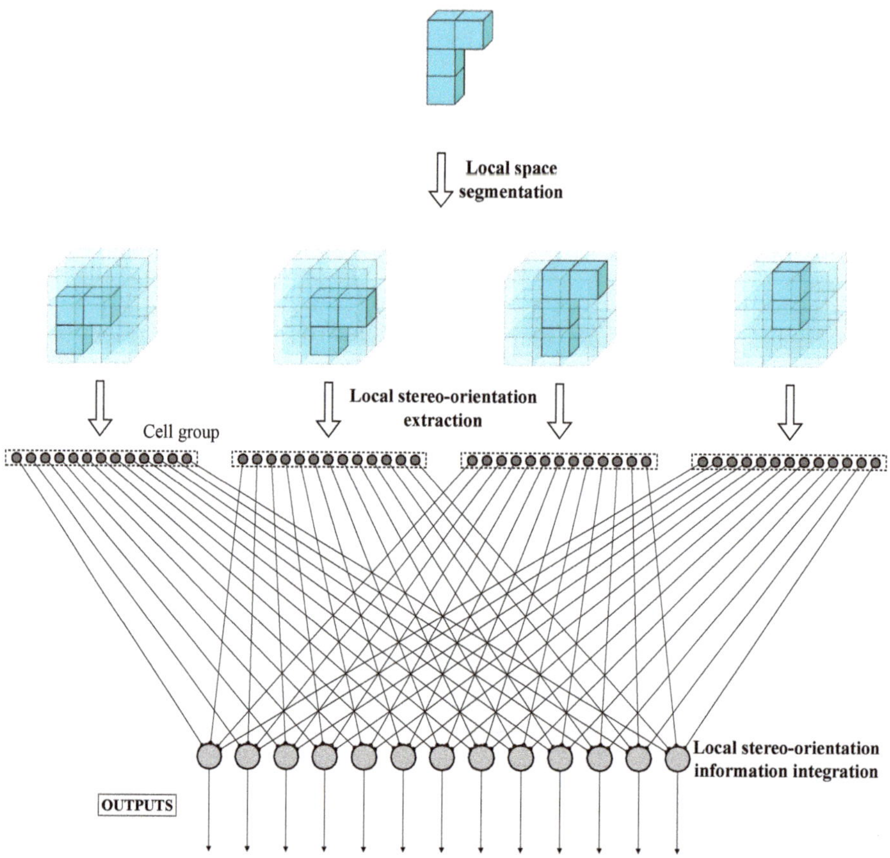

Figure 6. The overall process flow of AVS for stereo-orientation recognition.

3. Simulations and Results

3.1. Implementation Details

We mainly conducted three experiments based on the artificial visual system (AVS): Physiological visual simulation, stereo-orientation recognition, and 3D object classification. In this subsection, we give the implementation details of all experiments, including platform, implemented codes, evaluation criteria, and datasets.

Physiological visual simulation:

- Research question Validating the orientation selectivity of AVS
- Dataset 1 A grating dataset comprises 1600 frames, including 800 frames featuring gratings with four specific orientations and eight orthogonal moving directions, and 800 frames of blank data. Each 100 frames of drift motion was followed by 100 frames of static blank window simulating the resting duration after drifting.
- Dataset 2 A random dot dataset comprised 1300 3D data, and each consisted of 300 randomly positioned dots. Based on the original random dot dataset, randomly select several dots and perform specific forward motion and reverse motion to generate local orientation information. Through this process, five additional random dot datasets with local orientation fragments (number of moving dots: 1, 2, 3, 4, 5) were generated.
- Evaluation criteria The response intensity of complex stereo-orientation selective cells during grating drifting.

Stereo-orientation recognition

- Research question Validating the effectiveness of AVS in stereo-orientation information extraction and the superiority compared with deep models.
- Dataset An artificial object orientation dataset consists of various objects with specific orientations and scales (line segment or bar). Multiple noise test sets with varying intensities of background noise. Training set size: 15,600; Validation set size: 5200; Test set size: 5200.
- Evaluation criteria Detection accuracy on clean and noise stereo-orientation data.

3D object classification

- Research question Validating the effectiveness of AVS in enhancing deep model's robustness against noise.
- Dataset 3D-MNIST and 3D-Fashion-MNIST datasets derived from the original MNIST [10] and Fashion-MNIST image dataset [87].
- Evaluation criteria Detection accuracy on clean and noise 3D object datasets and the standard deviation of repeated simulations.

Abaltion study

- Research question Validating the contribution of each component in AVS to deep model performance enhancement.
- Dataset 3D-MNIST datasets.
- Evaluation criteria Detection accuracy of the deep model with different feature selection methods on clean 3D object datasets.

All experiments were implemented on the Apple M1 Max chips (Apple Inc., Cupertino, CA, USA) and NVIDIA GeForce RTX 3090 hardware environment (NVIDIA Corporation, Santa Clara, CA, USA). Implementation code related to this paper is available from the corresponding author upon request.

3.2. Physiological Visual Simulation

In the field of orientation selectivity research, drift gratings are commonly employed to stimulate visual neurons and record their responses, thereby determining their orientation selectivity. Similarly, we utilized drift gratings to initially validate the orientation selectivity of the artificial visual system (AVS) in our study. Figure 7 illustrates one frame of the drift grating data.

We employed AVS to detect the global orientation (0°, 45°, 90°, or 135°) within each frame of drift grating data and recorded the responses of complex stereo-orientation selective cells. The normalized activation outputs of four corresponding complex stereo-orientation selective cells are presented in Figure 8. As expected, for each specific orientation of the drift grating data, the respective selective cells exhibited vigorous activation throughout the entire motion period, while remaining in a resting state for non-preferred orientation. The simulation records are consistent with the previous results of physiological research [13,19,88] and initially demonstrate the feasibility of AVS in object orientation recognition.

Figure 7. One frame of drift grating stimuli data.

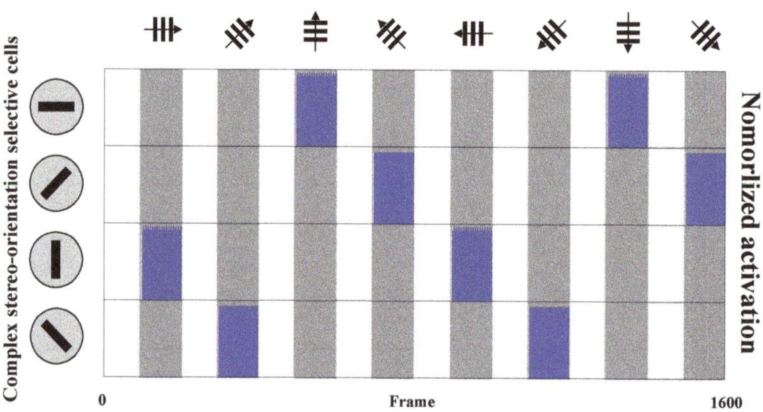

Figure 8. Activation records of complex stereo-orientation selective cells(0°, 45°, 90°, and 135°). The white area represents the rest state of stimuli, the gray area represents the moving state of stimuli, and the blue color denotes the neuron activation.

Moreover, the random dot pattern is also widely utilized for validating neuron activation in the field of stereovision research [89]. A random dot pattern consists of a large number of points that are spatially randomly distributed without any clear features. This allows the study of neuron responses to be as free from other cues as possible, thus enabling a pure investigation of the visual system's response to specific attributes. In this research, we select several dots in each 3D random dot data to perform specific forward and reverse motions and generate local stereo-orientation information for each data, as illustrated in Figure 9. Subsequently, we evaluated the AVS's sensitivity to oreintation information fragments within a large amount of random dots.

Summarized the detection accuracy of AVS on random dots datasets with different numbers of moving dots and illustrated in Figure 10. When the number of moving dots is zero, there is no orientation information, and the accuracy rate of AVS is 8.8%. Notably, even a single moving dot can provide effective orientation information for AVS (with 75.5% accuracy), and when the number of moving dots increased to three, AVS could achieve almost perfect performance. The results on the random dot pattern demonstrated that AVS is effective and robust in extracting orientation features under complex information disturbance.

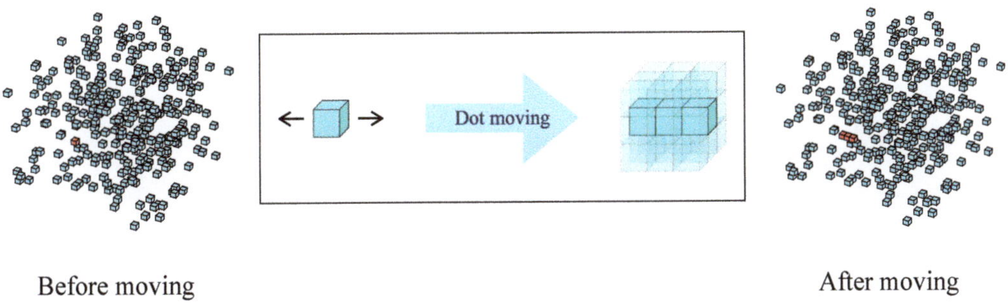

Figure 9. The generation process of oreintation information in random dots.

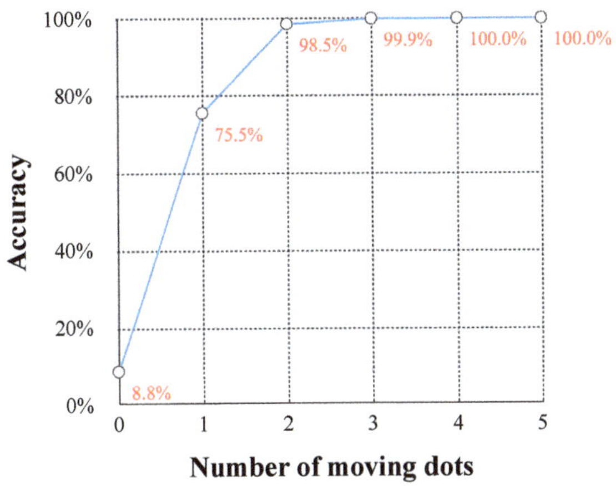

Figure 10. Acuuracy curve of AVS across different random dot datasets.

3.3. Stereo-Orientation Recognition

The effectiveness and robustness of AVS in stereo-orientation recognition are evaluated on the artificial object orientation datasets. Figure 11 presents several instances of both clean data and noise data.

The robustness of AVS against spatial noise jitter is evaluated using these noise test sets. We also employed multiple deep neural network models for object orientation recognition tasks. The voxel-based models were pre-loaded with ImageNet-based pre-trained parameters, and the point cloud-based models were trained without pre-trained parameters (see Table A1 in Appendix A for more details of training settings). The performances of various models are presented in Table 1.

Accuracy under Different Noise Intensities In the orientation detection task, all deep models exhibited good performance on clean data. The TransNeXt and PointMLP models even realized 100% accuracy. However, these deep models performed poorly under noise disturbance, and the accuracy dropped significantly as noise intensity increased. In contrast, AVS not only achieved perfect performance on clean data (100% accuracy), but it also demonstrated remarkable robustness to noise. AVS maintains 92.73% accuracy even at 5% intensity of noise.

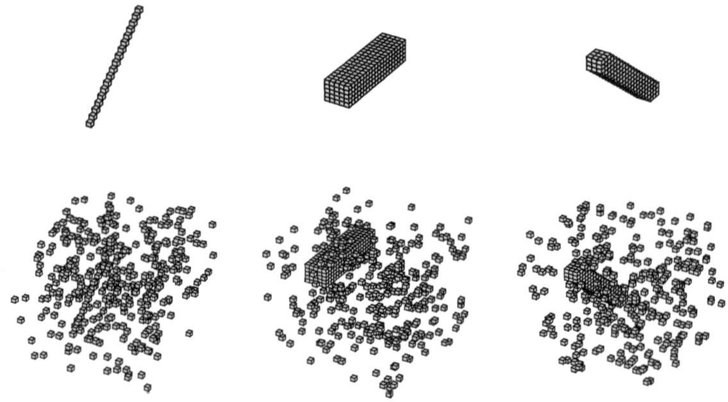

Figure 11. Instances of clean data and noise data.

Table 1. Detection accuracy on object orientation datasets.

Method	Noise Intensity						#Prams.	FLOPs
	0%	1%	2%	3%	4%	5%		
ConvNeXt [90]	99.96%	34.83%	26.52%	23.58%	21.65%	21.10%	87.61 M	3890 M
UniRepLKNet [91]	99.04%	19.90%	17.54%	15.98%	16.29%	14.60%	62.35 M	60 M
SwinV2 [92]	95.25%	23.08%	20.94%	20.27%	19.88%	19.34%	49.02 M	210 M
TransNeXt [93]	**100%**	36.71%	34.71%	30.92%	28.63%	25.63%	12.44 M	2630 M
PointNet [94]	99.85%	7.62%	7.69%	7.69%	7.69%	7.69%	3.46 M	230 M
PointMLP [95]	**100%**	8.25%	7.58%	7.58%	7.65%	7.67%	13.23 M	7890 M
AVS	**100%**	**99.71%**	**98.31%**	**96.98%**	**94.98%**	**92.73%**	353	9.48 M

Model Complexity ConvNeXt and UniRepLKNet models possess the most parameters but did not attain an expected model performance. And despite the PointMLP model's highest FLOPs of 7890 M, its robustness to noise was virtually none. In contrast, AVS achieved excellent performance with only 353 parameters and FLOPs of 9.48 M.

In short, deep models require high computing resources but perform poorly on noise data, while the AVS achieves the optimal performance with the lowest computational cost. AVS is feasible, efficient, and robust in stereo-orientation information extraction.

3.4. 3D Object Classification

We further extended the application of AVS for 3D object classification tasks. The implemented artificial visual system (AVS) based on the mechanism of stereo-orientation selectivity demonstrates high efficiency and remarkable robustness in orientation recognition tasks. Figure 12 illustrates the visualization of information processing within AVS. The stereo-orientation selective cells exclusively respond to the specific spatial orientation information, thereby the spatial elements like noise or some object elements without local orientation information would not elicit neuron responses. In contrast, deep neural network models consider all spatial information and allocate additional attention to spatial elements without stereo-orientation information, resulting in a significant decline in performance.

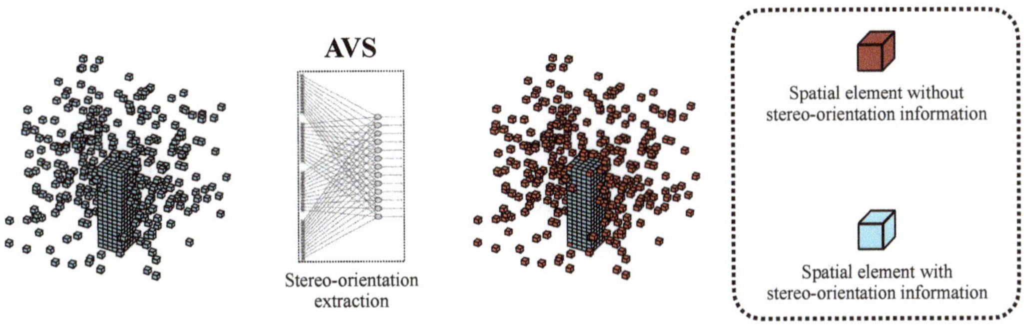

Figure 12. The visualization of information processing within AVS.

Similar to our explanation that specific spatial depth information can be organized into a specific stereo-orientation feature, we suggest that complex spatial features are the composition of multiple specific local stereo-orientation information. As illustrated in Figure 13, the digital number '4' can be exactly segmented into eight local stereo-orientation features. Considering the efficiency and robustness of AVS in extracting spatial features, we propose employing AVS as a pre-processing method in 3D object recognition tasks as a potential way to enhance the performance of deep neural network models.

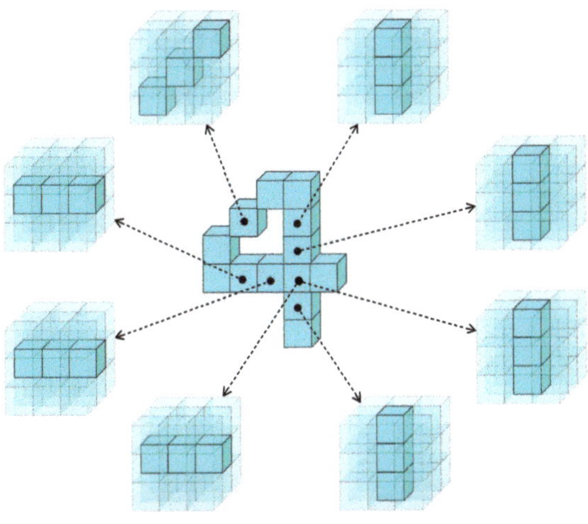

Figure 13. The digital number '4' and its composition of stereo-orientation information.

Based on AVS, we initially classify each spatial element as either positive or negative according to the presence or absence of local stereo-orientation information within the local space expanded from each element. The elements identified with orientation information are considered positive, while those without are classified as negative. The details of the separation process for a spatial element in a local space are described in Algorithm 1. Subsequently, we separate the negative and positive information and finally utilize the positive information as input for deep neural network models.

Algorithm 1 AVS for feature selection

Require: A spatial element p, local voxel array $voxel$
Ensure: 1 if orientation detected in $voxels$, otherwise 0
Initialize: 13 types of local stereo-orientation selective $cell_i$
$outs\_arr \leftarrow [0, 0, \ldots, 0]$ # length: 13
$\theta \leftarrow 10^4$ # cell response sensitivity
$k \leftarrow 2.5$ # cell activation threshold
$activation \leftarrow 0$ # initial cell activation value
for $i = 0$ **to** 12 **do**
  $cell_i \leftarrow orientation\ selective\ cell$ # create corresponding instance of cell class
  $activation \leftarrow cell_i(\theta, k, voxel)$ # cell activation computation
  $outs\_arr[i] \leftarrow activation$ #cell activation value: 1 or 0
end for
if $\sum_{i=0}^{12} outs\_arr[i] > 0$ **then**
  return 1 #positive information, preserve spatial element p
else
  return 0 #negative information, discard spatial element p
end if

The separation of effective and negative information from original spatial information is illustrated in Figure 14.

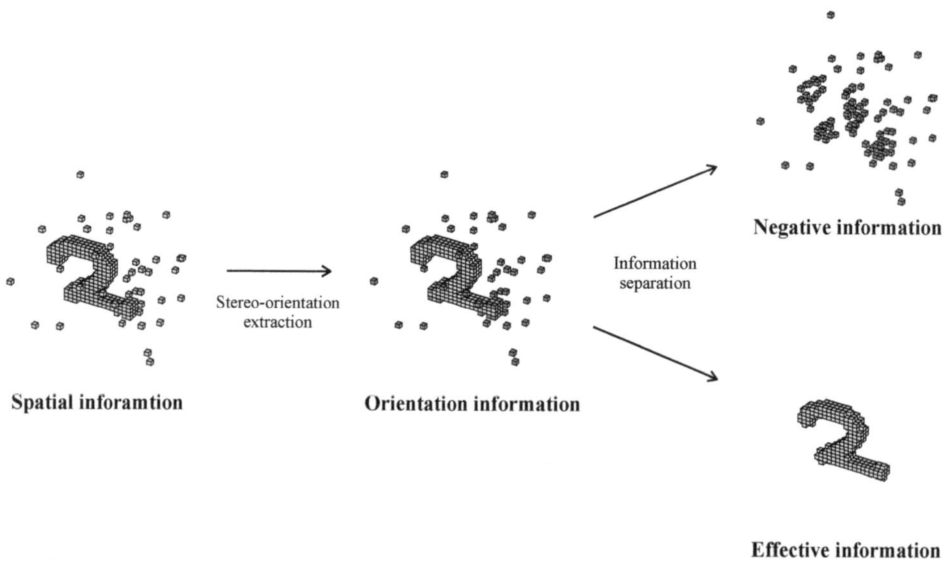

Figure 14. The separation of effective and negative information from original spatial information.

The reliability of AVS in recognizing effective object information and enhancing the deep model's stability and robustness is evaluated on 3D-MNIST and 3D-Fashion-MNIST datasets. TransNeXt and PointMLP, the two models performed well on orientation detection task, were employed as the baseline, and AVS-TransNeXt and AVS-PointMLP were implemented as comparisons to evaluate AVS's impact on deep model performance and robustness against noise. The models were trained on clean data and evaluated on both clean and noise test sets. The experiments were repeated 3 times under different random seed situations (see Table A2 in Appendix A for more details of training settings). The summarized results are presented in Tables 2 and 3.

Table 2. Detection accuracy of TransNeXt on 3D-MNIST and 3D-Fashion-MNIST datasets.

Dataset	Method	Clean [%]	Noise [%]	Overall [%]
MNIST	TransNext	**97.6 ± 0.1**	68.3 ± 16.3	73.1 ± 13.6
	AVS-TransNext	97.5 ± 0.2	**97.2 ± 0.2**	**97.2 ± 0.2**
Fashion-MNIST	TransNext	85.5 ± **0.4**	50.2 ± 10.6	56.1 ± 8.9
	AVS-TransNext	**85.7** ± 0.5	**84.4 ± 0.8**	**84.6 ± 0.8**

Table 3. Detection accuracy of PointMLP on 3D-MNIST and 3D-Fashion-MNIST datasets.

Dataset	Method	Clean [%]	Noise [%]	Overall [%]
MNIST	PointMLP	87.8 ± 1.3	12.6 ± **0.7**	25.1 ± **0.7**
	AVS-PointMLP	**91.0 ± 0.4**	**85.9** ± 1.0	**85.7** ± 0.9
Fashion-MNIST	PointMLP	73.7 ± 1.0	13.8 ± 1.2	23.8 ± 1.1
	AVS-PointMLP	**75.9 ± 0.1**	**75.5 ± 0.3**	**75.6 ± 0.3**

Results analysis:

- TransNeXt model performed well on clean data but with very poor stability to noise data. The results on noise data (68.3% ± 16.3 and 50.2% ± 10.6) exhibited large performance fluctuation. It is worth noting that anomalous situations occurred in different random seed settings, indicating that TransNeXt is sensitive to the initial parameters and susceptible to noise.
- AVS-TransNeXt model's performance on clean data is similar to the original TransNeXt. Concurrently, AVS-TransNeXt achieved remarkable noise robustness with small performance fluctuation (97.2% ± 0.2 and 84.4% ± 0.8).
- PointMLP model showed sensitivity and poor stability to noise data. Consequently, its performance was severely compromised, resulting in an accuracy of approximately 12% in 10-classification tasks.
- AVS-PointMLP model demonstrated 2% improved accuracy on clean data compared to the original PointMLP and significant improvement in accuracy and stability on noise data (85.9% ± 1.0 and 75.5% ± 0.3).

In summary, AVS significantly enhanced the baseline models' robustness and stability on noise data without performance degradation on clean data. AVS is demonstrated as an efficient feature preprocessing system that helps enhance the overall performance of deep models in 3D recognition tasks with minimal computation cost.

3.5. Ablation Study

To investigate the contribution of each component in AVS to deep model performance enhancement, we conducted an ablation study. When AVS was employed to enhance deep models' performance, the complex stereo-orientation selective cells were removed, and the depth selective cells were only responsible for accepting inputs. Accordingly, only the simple stereo-orientation selective cells performed computation in the information separation process. We can modify the separation rules by employing different combinations of simple stereo-orientation selective cells. We evaluated the impacts of different information separation rules based on local stereo-orientation on model performance. Based on the original separation rule, we defined three types of separation rules for positive information: multi-orientation, spatial elements with multiple types of local orientation information but not all; all orientation, the spatial elements with all types of local orientation information; single orientation, spatial elements with only one type of local orientation information. PointNet was employed as the baseline model. Employing the AVS with different separation rules to enhance the baseline model's performance and record the results as illustrated in Figure 15.

Because different types of information separation rules decided the amount of information extracted from original data, the three methods made different contributions to model performance enhancement. The 'single orientation' method demonstrated the most significant improvement in model performance, indicating that this method extracted more object features. And when considering all oreintation information states in local space, AVS could significantly improve the model's performance, suggesting that all simple stereo-orientation selective cells in AVS made a contribution.

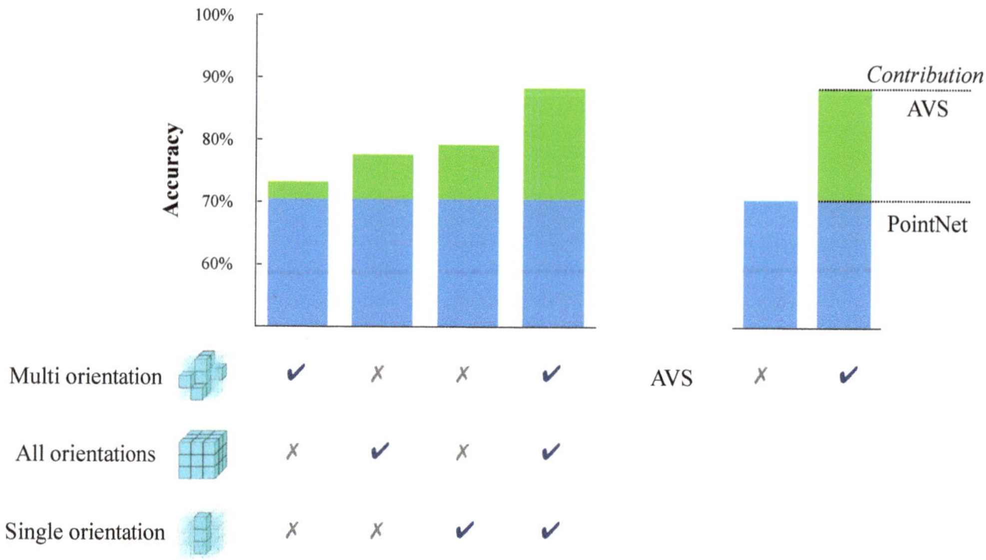

Figure 15. Ablation study on the feature selection process based on local stereo-orientation.

4. Discussion

Following the thought of local-to-global information aggregation in the Hubel-Wiesel model, we proposed a cascade system to quantitatively explain the generation mechanism of stereo-orientation selectivity and further innovatively developed an artificial visual system for object stereo-orientation selection. Based on the results of physiological visual simulation and object stereo-orientation detection, the proposed mechanism is feasible and effective. The AVS exhibited superiority on orientation extraction tasks compared with deep models. Despite their remarkable learning capability, deep models can learn enough knowledge to complete the detection task with clean data, but they were also affected by other potential spatial features, leading to weak attention to noise. The results demonstrate the weakness of deep models in robustness, which exhibits the gaps between the deep model and the real brain. The extended application of AVS in 3D object classification demonstrated that it can contribute to removing such a gap. After separating the spatial information into effective information (save) and negative information (discard) by AVS, the deep model significantly improved the robustness and stability.

It is noteworthy that the deep models employed for 3D object recognition tasks commonly involve a sampling process [96] like random sampling and farthest point sampling (FPS), which often sampling affects model performance. However, conventional sampling methods fail to eliminate the inclusion of noise information during the process. In voxel space, spatial elements of a 3D object are generally clustered together and have a higher probability of forming specific local stereo-orientation fragments, while background noise remains dissociated and independent. Therefore, we can address this issue by employing the AVS to separate noise information and object information. The spatial elements with local stereo-orientation information are identified as effective object information, while the spatial elements without local stereo-orientation information are noise and discarded. Figure 16 illustrates the point sampling effects with and without AVS processing. Notably, under noise situations, the random sampling and FPS methods sampled noise and object information simultaneously, leading to the model performance degradation in the subsequent learning phase. This is because the sampling methods are based on random selection or point coordinates. Consequently, the noise and object information have the same sampling probability under these sampling rules. After incorporating AVS processing, both sampling methods could obtain ideal sampling results, and all noise information was eliminated. Therefore, deep model performance enhancements described above benefit from the AVS completing the noise filtering to reduce the probability of sampling negative information.

We need to state that the current separation rule between negative and effective information remains incomplete. The variations in local stereo-orientation information content also decide the significance of spatial elements for the object recognition process. Furthermore, although AVS processing can effectively eliminate noise information, the subsequent sampling process still includes part of the stochastic process, and object information may be omitted. We can improve the feature extraction ability of AVS through flowing aspects:

- Incorporating the color perception mechanism into AVS to improve performance and generalizability in more complex scenes.
- Introducing a deep learning method to refine the information separation process and reduce information loss due to human design.
- Extending the application of AVS as a feature-based sampling method to avoid information loss during a stochastic process.

In future research, the enhanced AVS is anticipated to be utilized in medical image analysis, geological modeling, robotic vision, and industrial inspection. Furthermore, its fusion performance with other modalities, such as LiDAR and RGB images, will be explored.

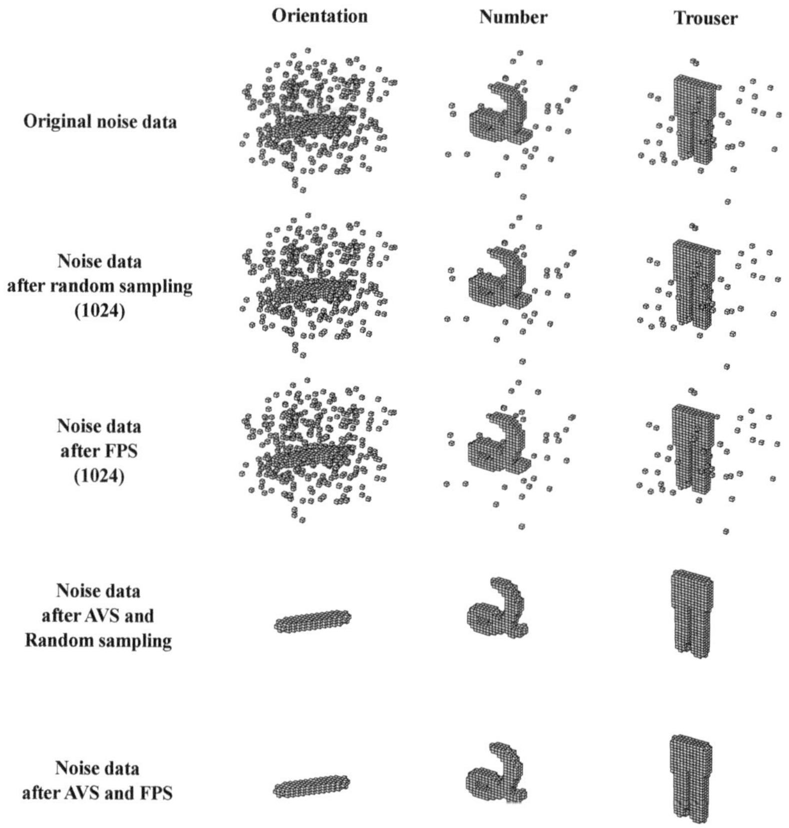

Figure 16. The point sampling effects with and without AVS processing.

5. Conclusions

In this paper, we innovatively explained the neural mechanism underlying stereo-orientation selectivity based on the Hubel-Wiesel Model, wherein the generation of stereo-orientation selectivity originates from the convergence and integration of multiple specific depth cues. We further developed an artificial visual system (AVS) for stereo-orientation recognition, comprising simple stereo-orientation selective cells that extract local spatial information and complex stereo-orientation selective cells responsible for integrating local orientation information to exhibit global selectivity. Simulation results demonstrate the effectiveness of AVS in feature extraction while maintaining robustness against spatial noise. Furthermore, we initially explore the integration of AVS with deep neural network models in 3D recognition tasks, which leads to significant enhancements in model robustness, performance, and stability. In conclusion, our research provides a reliable approach for feature extraction 3D object recognition tasks and helps explain how depth information is processed and integrated within the visual system to enable stereo vision.

Author Contributions: Conceptualization, B.L., Y.T. and Z.T.; methodology, B.L., Y.T. and Z.T.; simulation, B.L.; writing—original draft preparation, B.L.; writing—review and editing, B.L., Y.T. and Z.T. All authors have read and agreed to the published version of the manuscript.

Funding: This work was supported by JST, the establishment of university fellowships towards the creation of science technology innovation, Grant Number JPMJFS2116.

Institutional Review Board Statement: Not applicable.

Data Availability Statement: The data presented in this study are available on request from the corresponding author. The data are not publicly available due to data privacy regulations.

Conflicts of Interest: The authors declare no conflict of interest.

Abbreviations

The following abbreviations are used in this manuscript:

AVS	Artificial Vsiaul System
FPS	Farthest Point Sampling

Appendix A. Implementation Details of Deep Models

Table A1. Training settings for deep models on stereo-orientation recognition tasks.

(Pre-)Training Config	ConvNeXt-Base	UniRepLKNet-N	SwinV2-S	TransNeXt-Micro	PointNet	PointMLP
loss function	cross entropy	cross entropy	cross entropy	cross entropy	cross entropy	cross entropy
optimizer	AdamW	AdamW	AdamW	AdamW	AdamW	AdamW
base learning rate	1×10^{-4}	1×10^{-3}	1×10^{-4}	1×10^{-4}	1×10^{-4}	1×10^{-4}
weight decay	1×10^{-4}	1×10^{-4}	1×10^{-4}	1×10^{-4}	1×10^{-4}	1×10^{-4}
batch size	32	32	32	32	32	32
training epochs	30	30	30	10	30	10
learning rate schedule	exponential decay (0.8)	exponential decay (0.8)	exponential decay (0.8)	exponential decay (0.8)	exponential decay (0.8)	exponential decay (0.8)
number of sampling points	-	-	-	-	512 (random)	512 (FPS)
random seed	42	42	42	42	42	42

Table A2. Training settings for TransNeXt, PointMLP, and PointNet on 3D object classification tasks.

(Pre-)Training Config	TransNeXt-Micro	PointMLP	PointNet
loss function	cross entropy	cross entropy	cross entropy
optimizer	AdamW	AdamW	AdamW
base learning rate	1×10^{-5}	1×10^{-4}	1×10^{-4}
weight decay	1×10^{-4}	1×10^{-4}	1×10^{-4}
batch size	32	32	256
training epochs	10	10	30
learning rate schedule	exponential decay (0.8)	exponential decay (0.8)	exponential decay (0.8)
number of sampling points	-	1024 (FPS)	1024 (FPS)
random seeds	[7, 8, 42]	[7, 8, 42]	[7, 8, 42]

References

1. Zeki, S. *A Vision of the Brain*; Blackwell Scientific Publications: Cambridge, UK, 1993; p. 336.
2. Wandell, B.A. *Foundations of Vision*; Sinauer Associates: Sunderland, UK, 1995; p. 476.
3. Kandel, E.R.; Schwartz, J.H.; Jessell, T.M.; Siegelbaum, S.; Hudspeth, A.J.; Mack, S. (Eds.) *Principles of Neural Science*; McGraw-hill: New York, NY, USA, 2000; Volume 4.
4. Felleman, D.J.; Van Essen, D.C. Distributed hierarchical processing in the primate cerebral cortex. *Cereb. Cortex* **1991**, *1*, 1–47. [CrossRef] [PubMed]
5. Marr, D.; Hildreth, E. Theory of edge detection. *Proc. R. Soc. London. Ser. B. Biol. Sci.* **1980**, *207*, 187–217.
6. Landy, M.S.; Bergen, J.R. Texture segregation and orientation gradient. *Vis. Res.* **1991**, *31*, 679–691. [CrossRef]

7. Ferster, D.; Miller, K.D. Neural mechanisms of orientation selectivity in the visual cortex. *Annu. Rev. Neurosci.* **2000**, *23*, 441–471. [CrossRef] [PubMed]
8. Riesenhuber, M.; Poggio, T. Hierarchical models of object recognition in cortex. *Nat. Neurosci.* **1999**, *2*, 1019–1025. [CrossRef]
9. Fukushima, K. Neocognitron: A self-organizing neural network model for a mechanism of pattern recognition unaffected by shift in position. *Biol. Cybern.* **1980**, *36*, 193–202. [CrossRef]
10. LeCun, Y.; Bottou, L.; Bengio, Y.; Haffner, P. Gradient-based learning applied to document recognition. *Proc. IEEE* **1998**, *86*, 2278–2324. [CrossRef]
11. Camuffo, E.; Mari, D.; Milani, S. Recent advancements in learning algorithms for point clouds: An updated overview. *Sensors* **2022**, *22*, 1357. [CrossRef]
12. Ding, Z.; Sun, Y.; Xu, S.; Pan, Y.; Peng, Y.; Mao, Z. Recent advances and perspectives in deep learning techniques for 3D point cloud data processing. *Robotics* **2023**, *12*, 100. [CrossRef]
13. Hubel, D.H.; Wiesel, T.N. Receptive fields of single neurones in the cat's striate cortex. *J. Physiol.* **1959**, *148*, 574–591. [CrossRef]
14. Hubel, D.H.; Wiesel, T.N. Receptive fields, binocular interaction and functional architecture in the cat's visual cortex. *J. Physiol.* **1962**, *160*, 106. [CrossRef] [PubMed]
15. Hubel, D.H.; Wiesel, T.N. Receptive fields and functional architecture of monkey striate cortex. *J. Physiol.* **1968**, *195*, 215–243. [CrossRef]
16. Movshon, J.A.; Thompson, I.D.; Tolhurst, D.J. Receptive field organization of complex cells in the cat's striate cortex. *J. Physiol.* **1978**, *283*, 79–99. [CrossRef]
17. Hubel, D.H.; Wiesel, T.N. Receptive fields and functional architecture in two nonstriate visual areas (18 and 19) of the cat. *J. Neurophysiol.* **1965**, *28*, 229–289. [CrossRef]
18. Ferster, D.; Chung, S.; Wheat, H. Orientation selectivity of thalamic input to simple cells of cat visual cortex. *Nature* **1996**, *380*, 249–252. [CrossRef]
19. Hubel, D.H. Exploration of the primary visual cortex, 1955–78. *Nature* **1982**, *299*, 515–524. [CrossRef]
20. Alonso, J.M.; Martinez, L.M. Functional connectivity between simple cells and complex cells in cat striate cortex. *Nat. Neurosci.* **1998**, *1*, 395–403. [CrossRef]
21. Hubel, D.H.; Wiesel, T.N. Ferrier lecture-Functional architecture of macaque monkey visual cortex. *Proc. R. Soc. London. Ser. B. Biol. Sci.* **1977**, *198*, 1–59.
22. Liu, X.; Robinson, P.A. Analytic model for feature maps in the primary visual cortex. *Front. Comput. Neurosci.* **2022**, *16*, 659316. [CrossRef]
23. Leibo, J.Z.; Cornebise, J.; Gómez, S.; Hassabis, D. Approximate hubel-wiesel modules and the data structures of neural computation. *arXiv* **2015**, arXiv:1512.08457.
24. Rossi, L.F.; Harris, K.D.; Carandini, M. Spatial connectivity matches direction selectivity in visual cortex. *Nature* **2020**, *588*, 648–652. [CrossRef] [PubMed]
25. Lindeberg, T. Orientation selectivity of affine Gaussian derivative based receptive fields. *arXiv* **2023**, arXiv:2304.11920v9.
26. Wang, W.; Covi, E.; Milozzi, A.; Farronato, M.; Ricci, S.; Sbandati, C.; Pedretti, G.; Ielmini, D. Neuromorphic motion detection and orientation selectivity by volatile resistive switching memories. *Adv. Intell. Syst.* **2021**, *3*, 2000224. [CrossRef]
27. Han, W.J.; Han, I.S. Bio-inspired computing of vision—Fuzzy and neuromorphic processing. In Proceedings of the 2011 IEEE International Conference on Fuzzy Systems (FUZZ-IEEE 2011), Taipei, Taiwan, 27–30 June 2011; pp. 747–750.
28. Gao, S.; Liu, X. Explaining Orientation Adaptation in V1 by Updating the State of a Spatial Model. *Front. Comput. Neurosci.* **2022**, *15*, 759254. [CrossRef]
29. Yang, K.F.; Gao, S.B.; Guo, C.F.; Li, C.Y.; Li, Y.J. Boundary detection using double-opponency and spatial sparseness constraint. *IEEE Trans. Image Process.* **2015**, *24*, 2565–2578. [CrossRef]
30. Li, B.; Todo, Y.; Tang, Z.; Tang, C. The mechanism of orientation detection based on color-orientation jointly selective cells. *Knowl.-Based Syst.* **2022**, *254*, 109715. [CrossRef]
31. Grill-Spector, K.; Malach, R. The human visual cortex. *Annu. Rev. Neurosci.* **2004**, *27*, 649–677. [CrossRef]
32. Orban, G.A.; Van Essen, D.; Vanduffel, W. Comparative mapping of higher visual areas in monkeys and humans. *Trends Cogn. Sci.* **2004**, *8*, 315–324. [CrossRef]
33. Janssen, P.; Vogels, R.; Liu, Y.; Orban, G.A. Macaque inferior temporal neurons are selective for three-dimensional boundaries and surfaces. *J. Neurosci.* **2001**, *21*, 9419–9429. [CrossRef]
34. Hinkle, D.A.; Connor, C.E. Three-dimensional orientation tuning in macaque area V4. *Nat. Neurosci.* **2002**, *5*, 665–670. [CrossRef]
35. Orban, G.A. The extraction of 3D shape in the visual system of human and nonhuman primates. *Annu. Rev. Neurosci.* **2011**, *34*, 361–388. [CrossRef]
36. Yamane, Y.; Carlson, E.T.; Bowman, K.C.; Wang, Z.; Connor, C.E. A neural code for three-dimensional object shape in macaque inferotemporal cortex. *Nat. Neurosci.* **2008**, *11*, 1352–1360. [CrossRef]

37. Shikata, E.; Tanaka, Y.; Nakamura, H.; Taira, M.; Sakata, H. Selectivity of the parietal visual neurones in 3D orientation of surface of stereoscopic stimuli. *Neuroreport* **1996**, *7*, 2389–2394. [CrossRef] [PubMed]
38. Srinath, R.; Emonds, A.; Wang, Q.; Lempel, A.A.; Dunn-Weiss, E.; Connor, C.E.; Nielsen, K.J. Early emergence of solid shape coding in natural and deep network vision. *Curr. Biol.* **2021**, *31*, 51–65. [CrossRef]
39. DeAngelis, G.C.; Cumming, B.G.; Newsome, W.T. Cortical area MT and the perception of stereoscopic depth. *Nature* **1998**, *394*, 677–680. [CrossRef] [PubMed]
40. Uka, T.; Tanabe, S.; Watanabe, M.; Fujita, I. Neural correlates of fine depth discrimination in monkey inferior temporal cortex. *J. Neurosci.* **2005**, *25*, 10796–10802. [CrossRef] [PubMed]
41. Hinkle, D.A.; Connor, C.E. Disparity tuning in macaque area V4. *Neuroreport* **2001**, *12*, 365–369. [CrossRef] [PubMed]
42. Serre, T.; Oliva, A.; Poggio, T. A feedforward architecture accounts for rapid categorization. *Proc. Natl. Acad. Sci. USA* **2007**, *104*, 6424–6429. [CrossRef]
43. Pasupathy, A.; Connor, C.E. Shape representation in area V4: Position-specific tuning for boundary conformation. *J. Neurophysiol.* **2001**, *86*, 2505–2519. [CrossRef]
44. Poggio, G.F.; Poggio, T. The analysis of stereopsis. *Annu. Rev. Neurosci.* **1984**, *7*, 379–412. [CrossRef]
45. Qian, N. Binocular disparity and the perception of depth. *Neuron* **1997**, *18*, 359–368. [CrossRef]
46. Xu, X.; Cang, J.; Riecke, H. Development and binocular matching of orientation selectivity in visual cortex: A computational model. *J. Neurophysiol.* **2020**, *123*, 1305–1319. [CrossRef] [PubMed]
47. Yamins, D.L.; Hong, H.; Cadieu, C.; DiCarlo, J.J. Hierarchical modular optimization of convolutional networks achieves representations similar to macaque IT and human ventral stream. *Adv. Neural Inf. Process. Syst.* **2013**, *2*, 3093–3101 .
48. Yamins, D.L.; DiCarlo, J.J. Using goal-driven deep learning models to understand sensory cortex. *Nat. Neurosci.* **2016**, *19*, 356–365. [CrossRef]
49. Tanabe, S.; Fu, J.; Cang, J. Strong tuning for stereoscopic depth indicates orientation-specific recurrent circuitry in tree shrew V1. *Curr. Biol.* **2022**, *32*, 5274–5284. [CrossRef]
50. Cang, J.; Fu, J.; Tanabe, S. Neural circuits for binocular vision: Ocular dominance, interocular matching, and disparity selectivity. *Front. Neural Circuits* **2023**, *17*, 1084027. [CrossRef] [PubMed]
51. Read, J.C. Binocular vision and stereopsis across the animal kingdom. *Annu. Rev. Vis. Sci.* **2021**, *7*, 389–415. [CrossRef] [PubMed]
52. La Chioma, A.; Bonhoeffer, T.; Hübener, M. Disparity sensitivity and binocular integration in mouse visual cortex areas. *J. Neurosci.* **2020**, *40*, 8883–8899. [CrossRef] [PubMed]
53. Duan, R.; Paudel, D.P.; Fu, C.; Lu, P. Stereo orientation prior for UAV robust and accurate visual odometry. *IEEE/ASME Trans. Mechatronics* **2022**, *27*, 3440–3450. [CrossRef]
54. Zhang, J.; Liu, Z.; Gao, Y.; Zhang, G. Robust method for measuring the position and orientation of drogue based on stereo vision. *IEEE Trans. Ind. Electron.* **2020**, *68*, 4298–4308. [CrossRef]
55. Islam, A.; Asikuzzaman, M.; Khyam, M.O.; Noor-A-Rahim, M.; Pickering, M.R. Stereo vision-based 3D positioning and tracking. *IEEE Access* **2020**, *8*, 138771–138787. [CrossRef]
56. Nagata, F.; Miki, K.; Imahashi, Y.; Nakashima, K.; Tokuno, K.; Otsuka, A.; Watanabe, K.; Habib, M. Orientation detection using a CNN designed by transfer learning of AlexNet. In Proceedings of the 8th IIAE International Conference on Industrial Application Engineering, Shimane, Japan, 26–30 March 2020; Volume 5, pp. 26–30.
57. Lin, C.Y.; Setiawan, E. Object orientation recognition based on SIFT and SVM by using stereo camera. In Proceedings of the 2008 IEEE International Conference on Robotics and Biomimetics, Bangkok, Thailand, 22–25 February 2009; pp. 1371–1376.
58. Lowe, D.G. Object recognition from local scale-invariant features. In Proceedings of the Seventh IEEE International Conference on Computer Vision, Corfu, Greece, 20–25 September 1999; Volume 2, pp. 1150–1157.
59. Lourakis, M.; Zabulis, X. Model-based pose estimation for rigid objects. In *International Conference on Computer Vision Systems*; Springer: Berlin/Heidelberg, Germany, 2013; pp. 83–92.
60. Yang, X.; Xu, Y.; Ye, L.; Zhao, X. Research progress on binocular stereo vision applications. *Laser Optoelectron. Prog.* **2023**, *60*, 0811010.
61. Liu, Y.; Yan, Z.; Tan, J.; Li, Y. Multi-purpose oriented single nighttime image haze removal based on unified variational retinex model. *IEEE Trans. Circuits Syst. Video Technol.* **2022**, *33*, 1643–1657. [CrossRef]
62. Tahir, N.U.A.; Zhang, Z.; Asim, M.; Chen, J.; ELAffendi, M. Object detection in autonomous vehicles under adverse weather: A review of traditional and deep learning approaches. *Algorithms* **2024**, *17*, 103. [CrossRef]
63. Yang, L.; Cao, J.; Chen, W.; Wang, H.; He, L. An efficient multi-scale transformer for satellite image dehazing. *Expert Syst.* **2024**, *41*, e13575. [CrossRef]
64. Zhu, Y.; Wang, T.; Fu, X.; Yang, X.; Guo, X.; Dai, J.; Qiao, Y.; Hu, X. Learning weather-general and weather-specific features for image restoration under multiple adverse weather conditions. In Proceedings of the IEEE/CVF Conference on Computer Vision and Pattern Recognition, Vancouver, BC, Canada, 17–24 June 2023; pp. 21747–21758.

65. Li, R.; Tan, R.T.; Cheong, L.F. All in one bad weather removal using architectural search. In Proceedings of the IEEE/CVF Conference on Computer Vision and Pattern Recognition, Seattle, WA, USA, 14–19 June 2020; pp. 3175–3185.
66. Peng, D.; Lei, Y.; Hayat, M.; Guo, Y.; Li, W. Semantic-aware domain generalized segmentation. In Proceedings of the IEEE/CVF Conference on Computer Vision and Pattern Recognition, New Orleans, LA, USA, 18–24 June 2022; pp. 2594–2605.
67. Kennerley, M.; Wang, J.G.; Veeravalli, B.; Tan, R.T. 2pcnet: Two-phase consistency training for day-to-night unsupervised domain adaptive object detection. In Proceedings of the IEEE/CVF Conference on Computer Vision and Pattern Recognition, Vancouver, BC, Canada, 17–24 June 2023; pp. 11484–11493.
68. Gupta, H.; Kotlyar, O.; Andreasson, H.; Lilienthal, A.J. Robust Object Detection in Challenging Weather Conditions. In Proceedings of the IEEE/CVF Winter Conference on Applications of Computer Vision, Waikoloa, HI, USA, 3–8 January 2024; pp. 7523–7532.
69. Nabi, M.T.; Ali, S.; Mahmood, Z.; Khan, M.A.; Alsenan, S. A self-supervised deep-driven model for automatic weather classification from remote sensing images. *Int. J. Remote Sens.* **2024**, 1–26. [CrossRef]
70. Li, X.; Liu, W.; Li, X.; Tan, H. Physical Perception Network and an All-weather Multi-modality Benchmark for Adverse Weather Image Fusion. *arXiv* **2024**, arXiv:2402.02090.
71. Bijelic, M.; Gruber, T.; Mannan, F.; Kraus, F.; Ritter, W.; Dietmayer, K.; Heide, F. Seeing through fog without seeing fog: Deep multimodal sensor fusion in unseen adverse weather. In Proceedings of the IEEE/CVF Conference on Computer Vision and Pattern Recognition, Seattle, WA, USA, 13–19 June 2020; pp. 11682–11692.
72. Vincent, J.F. Biomimetics—a review. *Proc. Inst. Mech. Eng. Part H J. Eng. Med.* **2009**, *223*, 919–939. [CrossRef]
73. Zhang, Z.; Wang, Q.; Zhang, S. Review of computational fluid dynamics analysis in biomimetic applications for underwater vehicles. *Biomimetics* **2024**, *9*, 79. [CrossRef] [PubMed]
74. Peng, Y.; Yang, X.; Li, D.; Ma, Z.; Liu, Z.; Bai, X.; Mao, Z. Predicting flow status of a flexible rectifier using cognitive computing. *Expert Syst. Appl.* **2024**, *264*, 125878. [CrossRef]
75. Peng, Y.; He, M.; Hu, F.; Mao, Z.; Huang, X.; Ding, J. Predictive Modeling of Flexible EHD Pumps using Kolmogorov-Arnold Networks. *arXiv* **2024**, arXiv:2405.07488. [CrossRef]
76. Dapello, J.; Marques, T.; Schrimpf, M.; Geiger, F.; Cox, D.; DiCarlo, J.J. Simulating a primary visual cortex at the front of CNNs improves robustness to image perturbations. *Adv. Neural Inf. Process. Syst.* **2020**, *33*, 13073–13087.
77. Bertoni, F.; Citti, G.; Sarti, A. LGN-CNN: A biologically inspired CNN architecture. *Neural Netw.* **2022**, *145*, 42–55. [CrossRef] [PubMed]
78. Janssen, P.; Vogels, R.; Orban, G.A. Three-dimensional shape coding in inferior temporal cortex. *Neuron* **2000**, *27*, 385–397. [CrossRef]
79. Todd, J.T.; Reichel, F.D. Ordinal structure in the visual perception and cognition of smoothly curved surfaces. *Psychol. Rev.* **1989**, *96*, 643. [CrossRef] [PubMed]
80. Koenderink, J.J.; Van Doorn, A.J. Surface shape and curvature scales. *Image Vis. Comput.* **1992**, *10*, 557–564. [CrossRef]
81. Tsutsui, K.I.; Taira, M.; Sakata, H. Neural mechanisms of three-dimensional vision. *Neurosci. Res.* **2005**, *51*, 221–229. [CrossRef] [PubMed]
82. Lazareva, O. Depth perception. In *Encyclopedia of Evolutionary Psychological Science*; Shackelford, T.K., Weekes-Shackelford, V.A., Eds.; Springer: Cham, Switzerland, 2021; pp. 1903–1908.
83. Aida, S.; Fukamachi, S.; Shimono, K. Depth perception of stereoscopic transparent stimuli with frame manipulation. *Sci. Rep.* **2024**, *14*, 6712. [CrossRef] [PubMed]
84. Park, S.; Shin, Y.G.; Ko, S.J. Contrast enhancement using sensitivity model-based sigmoid function. *IEEE Access* **2019**, *7*, 161573–161583. [CrossRef]
85. Hopfield, J.J. Neurons with graded response have collective computational properties like those of two-state neurons. *Proc. Natl. Acad. Sci. USA* **1984**, *81*, 3088–3092. [CrossRef]
86. Alonso, N.I. John Hopfield's Contributions to Neural Networks: A Detailed Mathematical Exploration. *SSRN* **2024**. [CrossRef]
87. Xiao, H.; Rasul, K.; Vollgraf, R. Fashion-mnist: A novel image dataset for benchmarking machine learning algorithms. *arXiv* **2017**, arXiv:1708.07747.
88. Ohki, K.; Chung, S.; Ch'ng, Y.H.; Kara, P.; Reid, R.C. Functional imaging with cellular resolution reveals precise micro-architecture in visual cortex. *Nature* **2005**, *433*, 597–603. [CrossRef] [PubMed]
89. Dev, P. Perception of depth surfaces in random-dot stereograms: A neural model. *Int. J. Man-Mach. Stud.* **1975**, *7*, 511–528. [CrossRef]
90. Liu, Z.; Mao, H.; Wu, C.Y.; Feichtenhofer, C.; Darrell, T.; Xie, S. A convnet for the 2020s. In Proceedings of the IEEE/CVF Conference on Computer Vision and Pattern Recognition, New Orleans, LA, USA, 18–24 June 2022; pp. 11976–11986.
91. Ding, X.; Zhang, Y.; Ge, Y.; Zhao, S.; Song, L.; Yue, X.; Shan, Y. UniRepLKNet: A Universal Perception Large-Kernel ConvNet for Audio Video Point Cloud Time-Series and Image Recognition. In Proceedings of the IEEE/CVF Conference on Computer Vision and Pattern Recognition, Seattle, WA, USA, 16–22 June 2024; pp. 5513–5524.

92. Liu, Z.; Hu, H.; Lin, Y.; Yao, Z.; Xie, Z.; Wei, Y.; Ning, J.; Cao, Y.; Zhang, Z.; Dong, L.; et al. Swin transformer v2: Scaling up capacity and resolution. In Proceedings of the IEEE/CVF Conference on Computer Vision and Pattern Recognition, New Orleans, LA, USA, 18–24 June 2022; pp. 12009–12019.
93. Shi, D. TransNeXt: Robust Foveal Visual Perception for Vision Transformers. In Proceedings of the IEEE/CVF Conference on Computer Vision and Pattern Recognition, Seattle, WA, USA, 16–24 June 2024; pp. 17773–17783.
94. Qi, C.R.; Su, H.; Mo, K.; Guibas, L.J. Pointnet: Deep learning on point sets for 3d classification and segmentation. In Proceedings of the IEEE Conference on Computer Vision and Pattern Recognition, Honolulu, HI, USA, 21–26 July 2017; pp. 652–660.
95. Ma, X.; Qin, C.; You, H.; Ran, H.; Fu, Y. Rethinking network design and local geometry in point cloud: A simple residual MLP framework. *arXiv* **2022**, arXiv:2202.07123.
96. Zhang, H.; Wang, C.; Tian, S.; Lu, B.; Zhang, L.; Ning, X.; Bai, X. Deep learning-based 3D point cloud classification: A systematic survey and outlook. *Displays* **2023**, *79*, 102456. [CrossRef]

biomimetics

MDPI

Article

Adaptive Neural Network Robust Control of FOG with Output Constraints

Shangbo Liu [1,*,†], Baowang Lian [1,†], Jiajun Ma [2,†], Xiaokun Ding [3,†] and Haiyan Li [4,†]

1 School of Electronics and Information, Northwestern Polytechnical University, 127 West Youyi Road, Beilin District, Xi'an 710072, China; bwlian@nwpu.edu.cn
2 Electronic and College of Big Data and Communication Engineering Information Engineering, Guizhou University, Guiyang 550025, China; jjma3@gzu.edu.cn
3 AVIC Flight Automatic Control Research Institute, Northwestern Polytechnical University, 127 West Youyi Road, Beilin District, Xi'an 710072, China; dingxiaokun@aliyun.com
4 School of Life Science and Technology, Changchun University of Science and Technology, Changchun 130022, China; lihaiyan@cust.edu.cn
* Correspondence: liusb.438@163.com
† These authors contributed equally to this work.

Abstract: In this work, an adaptive robust control method based on Radial Basis Function Neural Network (RBFNN) is proposed. Inspired by the local response characteristics of biological neurons, this method can reduce the influence of nonlinear errors and unknown perturbations in the extreme working conditions of the aircraft, such as high dynamics and strong vibration, so as to achieve high tracking accuracy. In this method, the dynamic model of the nonlinear error of the fiber optic gyroscope is proposed, and then the unknown external interference observer is designed for the system to realize the estimation of the unknown disturbances. The controller design method combines the design of the adaptive law outside the finite approximation domain of the achievable condition design of the sliding mode surface, and adjusts the controller parameters online according to the conditions satisfied by the real-time error state, breaking through the limitation of the finite approximation domain of the traditional neural network. In the finite approximation domain, an online adaptive controller is constructed by using the universal approximation ability of RBFNN, so as to enhance the robustness to nonlinear errors and external disturbances. By designing the output constraint mechanism, the dynamic stability of the system is further guaranteed under the constraints, and finally its effectiveness is verified by simulation analysis, which provides a new solution for high-precision inertial navigation.

Keywords: fiber optic gyroscope; neural network; adaptive control; output constraints

1. Introduction

The optical fiber gyroscope (FOG) is an interferometric angular rate sensor based on the Sagnac effect, constructed from a ring of optical fibers. When the optical fiber ring rotates around its normal, two coherent light beams traveling in opposite directions within the ring will produce a phase difference proportional to the angular rotational rate. By detecting changes in the interference intensity caused by this phase difference, the angular rate can be measured. The FOG is the second generation of optical gyroscope after laser gyroscope, which has become the core device of high-precision navigation system for various aircraft because of its advantages of fast start-up, high precision, high bandwidth, all-solid-state, and strong environmental adaptability, and has been widely used in sea, land, air, space and other fields [1–5], and its performance directly affects the navigation accuracy of aircraft. High-precision FOGs typically adopt a fully digital closed-loop processing scheme, using Y-waveguide integrated optical phase modulators as core components (multifunction integrated optic circuit, MIOC), and introducing a feedback control loop to compensate in real time for the phase difference caused by rotation around zero using the Y-waveguide, thereby forming a closed-loop control system. This scheme can significantly improve the dynamic range and nonlinearity of the scale factor of the fiber optic gyroscope. The closed-loop fiber optic gyroscope relies on the performance of the control system to stably track the input angular velocity, and the performance of the control system directly affects the dynamic output accuracy of the FOG. However, with the rapid development of aviation and aerospace technology, the emergence of new aircraft such as super-aircraft and the continuous improvement of the flight ability of the aircraft, the FOG will work in a long-term extreme dynamic environment accompanied by high dynamics, strong vibration and impact, and the fiber optic gyroscope will be subject to nonlinear errors and disturbance interference, and the tracking performance of its closed-loop control system will be greatly limited, resulting in the aggravation of the dynamic output error of the FOG, which directly affects the inertial navigation error and causes catastrophic consequences.

In order to improve the closed-loop control performance of FOGs in harsh flight environments, all parties have made this effort. In [6], a high-precision real-time detection and closed-loop control method of FOG loop gain was proposed, and the closed-loop automatic control loop of the gain was used to achieve the stable maintenance of the loop gain at the preset reference optimal value. In order to eliminate the steady-state error and appropriately increase the system bandwidth in [7], the proposed digital controller adds a proportional link to the original integration link to improve the dynamic characteristics of the FOG. In [8], in order to eliminate the steady-state error, an integral link and a moving average filter are added to the digital controller, and a proportional integral derivative (PID) control algorithm is introduced to improve the system dynamics. The research in the above literature shows that the traditional PID controller has achieved certain improvements in eliminating steady-state errors, suppressing deviation changes, and improving response speed. However, in the process of FOG, the vibration or other interference factors of the instrument will cause the characteristic parameters or results of the FOG to change, and the control strategy of using the fixed parameters of the traditional PID controller often leads to poor control effect. Therefore, in order to cope with the special environmental changes, many scholars have adopted fuzzy logic control to further optimize and improve the parameters of the PID controller to improve the stability and robustness of the controller, so as to better meet the control needs of the gyroscope in the process of operation and ensure the stability and reliability of the control effect, as described in. A novel F-PID composite controller combined with fuzzy control was proposed in [8]. The deviation value of input and output and the change rate of deviation are input language variables, and the parameters of PID are output language variables, so as to realize online adjustment of PID parameters and improve the static and dynamic characteristics of FOGs. However, the number of fuzzy rules involved in the design of the controller is large, and the design of the fuzzy rules and membership function is entirely based on experience, so the design of the controller is relatively difficult. In [9], the improved BP neural network was used to adjust and optimize the parameters of the common PID, and the influence of the previous output value of the PID on the output value of the PID was suppressed through active series correction, which solved the problems of local extreme value and slow convergence speed of the BP neural network. In [10], the improved gray wolf algorithm is used to tune the traditional PID parameters, and the Euclidean distance rate of change is used to dynamically adjust the convergence factor to balance the global search ability of the algorithm. The dynamic adaptive weight factor is introduced to improve the optimization speed and accuracy of the algorithm. Although the controller in Refs. [8–10] can achieve the desired control effect, the optimization process is complex, computationally intensive, and time-consuming. In [11], a fuzzy PID controller based on the gravitational search algorithm is used to realize the closed-loop control of FOG, which has advantages in dynamic performance optimization, but the algorithm is optimized globally through the gravitational search algorithm, but its parameter adjustment requires multiple trial and error, and after coupling with the parameters of fuzzy PID, the overall optimization process takes a long time, which limits its practicability. In [12,13], a hybrid control system is designed by combining the advantages of fuzzy logic system and neural network, but due to its high interpretability and learning ability, the computational complexity is high, which leads to an increase in the amount of online computation, and it is difficult to use it in a system with high real-time performance.

In recent years, with the maturity of artificial neural network technology, Radial Basis Function Neural Network (RBFNN) is a kind of feedforward neural network based on local approximation theory, which has attracted great attention from researchers in various fields due to its simple structure, strong nonlinear mapping ability and fast learning speed. In [14], RBFNN was used to compensate for the bias error of the laser gyroscope, and the approximation effect of the RBFNN algorithm on the nonlinear error model was verified, and the improved RBFNN algorithm had the best performance. In [15–19], the RBFNN dynamic identification of the dynamic characteristics of MEMS gyroscopes and the approximation of external disturbance are used to ensure that the control system can reach the sliding surface from any initial state and converge to the equilibrium point in a limited time by introducing a sliding mode controller, reducing the number of parameters and avoiding the problem of parameter expansion. Finally, the simulation results show that in the presence of model uncertainty and external interference, RBFNN optimization control can reduce the input flutter and improve the timeliness and effectiveness of tracking.

The design inspiration comes from the local receptive fields and global information integration mechanisms in biological neural systems, which are inherently consistent with the multi-channel parallel processing characteristics of insect compound eyes in biologically inspired visual systems [20–22]. This simulates the ability of biological neurons to locally approximate nonlinear disturbances, achieving autonomous parameter optimization of the FOG in high-dynamic environments and carries out the simulation verification of the related control algorithms, so that the FOG control system can quickly realize high-precision closed-loop control under nonlinear error and uncertain disturbances.

2. Control System Model of FOG

The structure of the digital closed-loop FOG is shown in Figure 1. The optical section includes the light source, coupler, phase modulator, optical fiber loop, and photodetector, while the electrical section comprises a preamplifier, analog-to-digital converter (ADC), logic processor, digital-to-analog converter (DAC), and its output buffer amplifier, among others. The light source, coupler, phase modulator, and optical fiber loop form the Sagnac interferometer. When the fiber optic loop rotates, a non-reciprocal phase shift is generated between the counterpropagating light waves in the interferometer. The phase shift Φ_s is proportional to the angular velocity Ω, with the scale factor given by $K_s = \approx 2pi\frac{LD}{\lambda c}$, where L is the length of the optical fiber loop, D is the average diameter of the fiber loop, λ is the average wavelength of the light source, and c is the speed of light in a vacuum.

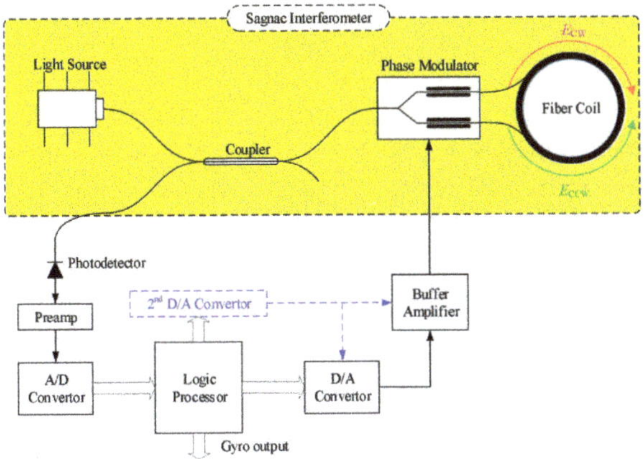

Figure 1. Structure diagram of the FOG System.

In the digital closed-loop FOG, The light emitted from the light source is split into two beams traveling in opposite directions after passing through the coupler and Y waveguide. The Y waveguide integrates beam-splitting, polarization, and phase modulation functions, and is therefore also known as a multifunction integrated optic circuit (MIOC). These two beams propagate in clockwise and counterclockwise directions within the fiber ring and interfere at the beam-splitting port of the Y waveguide. The spatial rotation angular velocity induces a phase difference between the two beams, which leads to changes in the interference light intensity. This interference signal is then output through the coupler to the photodetector. By processing the detected light intensity signal, the corresponding angular velocity can be extracted. To ensure stable operation of the fiber optic gyroscope at the optimal response point and to improve the linearity of the scale factor and the anti-interference capability, a closed-loop feedback mechanism is implemented. Leveraging the phase modulation functionality of the MIOC, a compensating phase is applied in real time to cancel out the phase shift caused by rotation, thereby stabilizing the operating point at a zero differential phase state and enabling precise computation of the angular velocity based on the feedback signal. This achieves high linearity and low noise in angular rate measurement. Moreover, to maintain accuracy, the feedback loop must operate within its linear range. If the phase difference exceeds 2π, nonlinear errors may occur due to the system's inability to accurately track the actual phase change. To address this, periodic 2π-resetting is employed to keep the feedback signal within the linear region, thus enhancing both the accuracy and stability of the measurement.

The modulation phase generated by the Y-waveguide compensates for the phase induced by the Sagnac effect. Considering interference factors such as the nonlinear modulation effects of the Y-waveguide, the control model of the digital closed-loop FOG is given as follows:

$$\begin{cases} \ddot{\phi}_F + k_1 \cos(\phi_s - \phi_F)\dot{\phi}_F + k_2 \sin(\phi_s - \phi_F) + D(t) + u = 0 \\ y = \phi_F \end{cases} \tag{1}$$

where $\phi_F = [\phi_{F1}, \phi_{F2}, \phi_{F3}]^T$ represents the system state, ϕ_{F1}, ϕ_{F2}, ϕ_{F3} denote the location on XYZ axis, respectively, $\phi_s = [\phi_{s1}, \phi_{s2}, \phi_{s3}]^T$ denotes XYZ axis, respectively, $k_1 = \text{diag}\{k_{11}, k_{12}, k_{13}\}$, $k_2 = \text{diag}\{k_{21}, k_{22}, k_{23}\}$, k_{ij} the known signal vector of the system (with i = 1, 2 and j = 1, 2, 3), $D(t) = [D_1(t), D_2(t), D_3(t)]^T$ represents the unknown external disturbance, and u is the controller to be designed. $y = \phi_F = [y_1, y_2, y_3]^T$ represents the system output. The system (1) satisfies the following conditions:

Assumption 1. *The external disturbance* $D_j(t)$, *is assumed to be unknown but bounded, satisfying the condition,* $|D_i(t)| \leq \bar{D}_i$, *where* $\bar{y}_{dj}$ *is a known positive constant.*

Assumption 2. *The desired trajectory* $y_d = [y_{d1}, y_{d2}, y_{d3}]^T$ *is bounded and needs to be tracked by the location of the FOG system* (1). *This implies that the desired trajectory satisfies the condition* $|y_{dj}| \leq \bar{y}_{dj}$, *where* $\bar{y}_{dj}$ *is a known positive constant.*

Remark 1. *The disturbances and trajectories considered in this paper are indeed bounded. While unbounded disturbances pose a significant challenge due to the severe shocks they may introduce, which could potentially render the control ineffective, it is important to note that even in such cases, the functions remain bounded within a defined finite domain. This ensures that the system behavior can still be analyzed and controlled within practical limits.*

The above analysis indicates that under 2π-reset conditions, the modulation state lacks symmetry, resulting in strong coupling between the primary and secondary closed-loop controls in the digital closed-loop FOG. The symmetry of the modulation state is the fundamental condition for decoupling the primary closed-loop control and the secondary closed-loop control in the digital closed-loop FOG.

Design of the RBFNN Controller

The basic structure schematic diagram of the RBFNN controller is shown in Figure 2.

$$\begin{cases} \dot{x}_1 = x_2 = \dot{\phi}_F \\ \dot{x}_2 = \ddot{x}_1 = \ddot{\phi}_F = -k_1 \cos(\phi_s - x_1)x_2 - k_2 \sin(\phi_s - x_1) - u - D(t) \end{cases} \tag{2}$$

where the state vector is $x_1 = [x_{11}, x_{12}, x_{13}]^T$, $x_2 = [x_{21}, x_{22}, x_{23}]^T$, $h_{a1} \leq x_{11} \leq h_{a2}$, $h_{c1} \leq x_{13} \leq h_{c2}$, h_{a1}, h_{a2}, h_{b1}, h_{b2}, h_{c1}, and h_{c2} denote the bounded values. u is the controller signal in Figure 2 that will be designed.

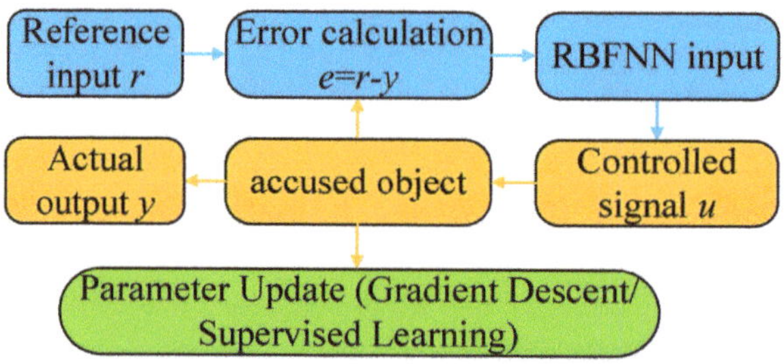

Figure 2. Structural schematic diagram of the RBFNN controller.

In Figure 2, the core parts of the RBFNN controller are the input layer, the controlled layer and the output layer; The controlled object module receives the controlled object u and outputs t parameters of the actual response signal y; In the parameter adjustment module, it adjusts the weights and basis function RBFNN based on error back propagation.

In this paper, let the signal $y_d = [y_{d1}, y_{d2}, y_{d3}]^T$ represent the desired signal, that is tracked by the position of state vector x_1 $\mu_1 = x_1 - y_d$, $\mu_2 = x_2 - \beta_1 = [\mu_{21}, \mu_{22}, \mu_{23}]^T$, where β_1 is the virtual controller to be designed. Then, it has $\dot{\mu}_1 = \dot{x}_1 - \dot{y}_d$, Take the Lyapunov function as:

$$V_1 = \frac{1}{2}\log\frac{h_{a1}^2}{h_{a1}^2 - \mu_{11}^2} + \frac{1}{2}\log\frac{h_{a2}^2}{h_{a2}^2 - \mu_{12}^2} + \frac{1}{2}\log\frac{h_{a3}^2}{h_{a3}^2 - \mu_{13}^2} \tag{3}$$

The derivative of function (3) is:

$$\begin{aligned} \dot{V}_1 &= \tfrac{\mu_{11}\dot{\mu}_{11}}{h_a^2-\mu_{11}^2} + \tfrac{\mu_{12}\dot{\mu}_{12}}{h_b^2-\mu_{12}^2} + \tfrac{\mu_{13}\dot{\mu}_{13}}{h_c^2-\mu_{13}^2} \\ &= [\tfrac{\mu_{11}}{h_a^2-\mu_{11}^2}, \tfrac{\mu_{12}}{h_b^2-\mu_{12}^2}, \tfrac{\mu_{13}}{h_c^2-\mu_{13}^2}](\mu_2 + \beta_1 - \dot{y}_d) \end{aligned} \tag{4}$$

Thus, the virtual controller can be designed as $\beta_1 = \dot{y}_d - r\mu_1$, where $r = \mathrm{diag}\{r_1, r_2, r_3\}$ r_1 denote positive real constant, so it obtains:

$$\begin{aligned} \dot{V}_1 = &-\tfrac{r_1\mu_{11}^2}{h_a^2-\mu_{11}^2} - \tfrac{r_2\mu_{12}^2}{h_b^2-\mu_{12}^2} - \tfrac{r_3\mu_{13}^2}{h_c^2-\mu_{13}^2} \\ &+\tfrac{\mu_{11}\mu_{21}}{h_a^2-\mu_{11}^2} + \tfrac{\mu_{12}\mu_{22}}{h_b^2-\mu_{12}^2} + \tfrac{\mu_{13}\mu_{23}}{h_c^2-\mu_{13}^2} \end{aligned} \tag{5}$$

Because $\dot{\mu}_2 = \dot{x}_2 - \dot{\beta}_1 = -k_1\cos(\phi_s - x_1)x_2 - k_2\sin(\phi_s - x_1) - u - D(t) - \dot{\beta}_1$, selecting the following Lyapunov function:

$$V_2 = V_1 + \frac{1}{2}\mu_2^T\mu_2 \tag{6}$$

$$\dot{V}_2 = \dot{V}_1 + \mu_2^T[-k_1\cos(\phi_s - x_1)x_2 - k_2\sin(\phi_s - x_1) - u - D(t) - \dot{\beta}_1] \tag{7}$$

Design the interference observer:

$$\hat{D} = P(t) + h(\mu_2)\mu_2 \tag{8}$$

$$\dot{P}(t) = h(\mu_2)[k_1\cos(\phi_s - x_1)x_2 + k_2\sin(\phi_s - x_1) + u + \hat{D} + \dot{\beta}_1] \tag{9}$$

where $P(t) = [P_1(t), P_2(t), P_3(t)]^T$, $h(\mu_2) = \mathrm{diag}\{h_1(\mu_2), h_2(\mu_2), h_3(\mu_2)\}$. Because $\tilde{D} = \hat{D} - D$, then it has:

$$\begin{aligned} \dot{\tilde{D}} &= \dot{D} - \dot{\hat{D}} = \dot{P}(t) + h(\mu_2)\dot{\mu}_2 - \dot{D} \\ &= h(\mu_2)[k_1\cos(\phi_s - x_1)x_2 + k_2\sin(\phi_s - x_1) + u + \hat{D} + \dot{\beta}_1] \\ &+h(\mu_2)[-k_1\cos(\phi_s - x_1)x_2 - k_2\sin(\phi_s - x_1) - u - D(t) - \dot{\beta}_1] - \dot{D} \\ &= h(\mu_2)\tilde{D} - \dot{D} \end{aligned} \tag{10}$$

From (10), it knows

$$\dot{\mu}_2 = -k_1\cos(\phi_s - x_1)x_2 - k_2\sin(\phi_s - x_1) - u - \hat{D} - \dot{\beta}_1 \tag{11}$$

Let the nonlinear function be $-k_1\cos(\phi_s - x_1)x_2 - k_2\sin(\phi_s - x_1) - \dot{\beta}_1 = \Delta$ At this point, the observers in Equations (8) and (9) contain unknown nonlinear terms, which cannot be directly applied. Therefore, it is necessary to redesign the observer based on the Radial Basis Function Neural Network (RBFNN). This paper considers the design of observers and controllers under two scenarios: when the states μ_2 are outside the RBFNN universal approximation domain and when they are within the approximation domain.

Theorem 1. *Under the conditions of Assumptions 1 and 2, for system* (1), *with the controller* $u = O_{3\times1}$, *disturbance observer* (12), *and adaptive laws* (13) *and* (14), *the extended state vector* $z = [\mu_2^T, \rho, \vartheta^T, \varepsilon^T, W_j^T]^T$ *will reach the sliding surface* s_1, s_2

$$\hat{D} = P(t) + h(\mu_2)\mu_2, \quad \dot{P}(t) = h(\mu_2)[\bar{\Delta} + u + \hat{D}] \tag{12}$$

The following form of adaptive law is designed:

$$\dot{\rho} = \begin{cases} \frac{1}{\alpha^2 \rho}(L + \Xi), & \|z\| \geq \|\mu_2\| > \alpha|\rho| \\ -\frac{1}{\alpha^2 \rho}(L + \Xi), & \|z\| \geq \alpha|\rho| > \|\mu_2\| \end{cases} \tag{13}$$

$$\dot{\hat{\vartheta}}_j = 0, \quad \dot{\hat{\varepsilon}}_j = 0, \quad \dot{\hat{W}}_j = O_{1\times n} \tag{14}$$

where $\Xi = \frac{|\mu_{11}|\cdot|\mu_{21}|}{|h_a^2 - \mu_{11}^2|} + \frac{|\mu_{12}|\cdot|\mu_{22}|}{|h_b^2 - \mu_{12}^2|} + \frac{|\mu_{13}|\cdot|\mu_{23}|}{|h_c^2 - \mu_{13}^2|} + \|\mu_2\|(\bar{\Delta} + |\hat{D}|)$.

Proof. Case 1: From the state vector $z = [\mu_2^T, \rho, \vartheta^T, \varepsilon^T, W_j^T]^T$, it is evident $\|z\| \geq \|\mu_2\|$ that when $\|z\| \geq \|\mu_2\| > \alpha|\rho|$ satisfies the given condition, the sliding surface can be chosen in the following form:

$$s_1 = V_1 + \frac{1}{2}\mu_2^T \mu_2 - \frac{1}{2}\alpha^2 \rho^2 + \sum_{j=1}^{3} \frac{1}{2\gamma_{1j}} \tilde{\vartheta}_j^2 + \sum_{j=1}^{3} \frac{1}{2\delta_{1j}} \tilde{\varepsilon}_j^2 + \sum_{j=1}^{3} \frac{1}{2\chi_{1j}} \tilde{W}_j^T \tilde{W}_j + \sum_{j=1}^{3} \frac{1}{2\tau_{1j}} \tilde{D}_j^2 \tag{15}$$

Order $\tilde{V} = \frac{1}{2}s_1^2$, there is:

$$\begin{aligned} \dot{\tilde{V}} = s_1\dot{s}_1 = s_1\{\dot{V}_1 + \mu_2^T[\Delta - u - \hat{D}] - \alpha^2\rho\dot{\rho} + \gamma_{1j}^{-1}\textstyle\sum_{j=1}^3 \tilde{\vartheta}_j\dot{\hat{\vartheta}}_j \\ + \delta_{1j}^{-1}\textstyle\sum_{j=1}^3 \tilde{\varepsilon}_j\dot{\hat{\varepsilon}}_j + \chi_{1j}^{-1}\textstyle\sum_{j=1}^3 \tilde{W}_j^T\dot{\hat{W}}_j + \tau_{1j}^{-1}\textstyle\sum_{j=1}^3 \tilde{D}_j\dot{\hat{D}}_j \end{aligned} \tag{16}$$

Using open-loop control $u = O_{3\times 1}$, the following conclusions are valid:

$$\begin{aligned} \dot{\tilde{V}} \leq s_1\{-\tfrac{r_1\mu_{11}^2}{h_a^2-\mu_{11}^2} - \tfrac{r_2\mu_{12}^2}{h_b^2-\mu_{12}^2} - \tfrac{r_3\mu_{13}^2}{h_c^2-\mu_{13}^2} + \tfrac{\mu_{11}\mu_{21}}{h_a^2-\mu_{11}^2} + \tfrac{\mu_{12}\mu_{222}}{h_b^2-\mu_{12}^2} + \tfrac{\mu_{13}\mu_{23}}{h_c^2-\mu_{13}^2} + \mu_2^T[\Delta - \hat{D}]\} \\ -\alpha^2\rho\dot{\rho} + \gamma_{1j}^{-1}\textstyle\sum_{j=1}^3 \tilde{\vartheta}_j\dot{\hat{\vartheta}}_j + \delta_{1j}^{-1}\textstyle\sum_{j=1}^3 \tilde{\varepsilon}_j\dot{\hat{\varepsilon}}_j + \chi_{1j}^{-1}\textstyle\sum_{j=1}^3 \tilde{W}_j^T\dot{\hat{W}}_j + \tau_{1j}^{-1}\textstyle\sum_{j=1}^3 \tilde{D}_j\dot{\hat{D}}_j\} \\ \leq s_1\{\tfrac{|\mu_{11}|\cdot|\mu_{21}|}{|h_a^2-\mu_{11}^2|} + \tfrac{|\mu_{12}|\cdot|\mu_{22}|}{|h_b^2-\mu_{12}^2|} + \tfrac{|\mu_{13}|\cdot|\mu_{23}|}{|h_c^2-\mu_{13}^2|} + \|\mu_2\|(\bar{\Delta} + |\hat{D}|) \\ -\alpha^2\rho\dot{\rho} + \gamma_{1j}^{-1}\textstyle\sum_{j=1}^3 \tilde{\vartheta}_j\dot{\hat{\vartheta}}_j + \delta_{1j}^{-1}\textstyle\sum_{j=1}^3 \tilde{\varepsilon}_j\dot{\hat{\varepsilon}}_j + \chi_{1j}^{-1}\textstyle\sum_{j=1}^3 \tilde{W}_j^T\dot{\hat{W}}_j + \tau_{1j}^{-1}\textstyle\sum_{j=1}^3 \tilde{D}_j\dot{\hat{D}}_j\} \\ \leq -s_1 L \end{aligned} \tag{17}$$

From conclusion (17), it can be inferred that the extended state z will reach the sliding surface s_1 within finite time.

Case 2: When $\|z\| \geq \alpha|\rho| \geq \|\mu_2\|$, satisfies the given condition, the sliding surface is designed as:

$$s_2 = V_1 + \frac{1}{2}\alpha^2\rho^2 - \frac{1}{2}\mu_2^T\mu_2 + \sum_{j=1}^{3}(\frac{1}{2\gamma_{1j}}\tilde{\vartheta}_j^2 + \frac{1}{2\delta_{1j}}\tilde{\varepsilon}_j^2 + \frac{1}{2\chi_{1j}}\tilde{W}_j^T\tilde{W}_j + \frac{1}{2\tau_{1j}}\tilde{D}_j^2) \tag{18}$$

Order $\tilde{V} = \frac{1}{2}s_1^2$, there is:

$$\begin{aligned} \dot{\tilde{V}} = s_1\dot{s}_1 = s_1\{\dot{V}_1 - \mu_2^T[\Delta - u - \hat{D}]\} \\ +\alpha^2\rho\dot{\rho} + \gamma_{1j}^{-1}\textstyle\sum_{j=1}^3 \tilde{\vartheta}_j\dot{\hat{\vartheta}}_j + \delta_{1j}^{-1}\textstyle\sum_{j=1}^3 \tilde{\varepsilon}_j\dot{\hat{\varepsilon}}_j + \chi_{1j}^{-1}\textstyle\sum_{j=1}^3 \tilde{W}_j^T\dot{\hat{W}}_j + \tau_{1j}^{-1}\textstyle\sum_{j=1}^3 \tilde{D}_j\dot{\hat{D}}_j \end{aligned} \tag{19}$$

Under open-loop control $u = O_{3\times 1}$, the following conclusion holds:

$$
\begin{aligned}
\dot{V} \le\ & s_2\{-\tfrac{r_1\mu_{11}^2}{h_a^2-\mu_{11}^2}-\tfrac{r_2\mu_{12}^2}{h_b^2-\mu_{12}^2}-\tfrac{r_3\mu_{13}^2}{h_c^2-\mu_{13}^2}+\tfrac{\mu_{11}\mu_{21}}{h_a^2-\mu_{11}^2}+\tfrac{\mu_{12}\mu_{22}}{h_b^2-\mu_{12}^2}+\tfrac{\mu_{13}\mu_{23}}{h_c^2-\mu_{13}^2}-\mu_2^{\mathrm{T}}(\Delta-\hat{D})\} \\
& +\alpha^2\rho\dot{\rho}+\gamma_{1j}^{-1}\textstyle\sum_{j=1}^3\tilde{\vartheta}_j\dot{\hat{\vartheta}}_j+\delta_{1j}^{-1}\sum_{j=1}^3\tilde{\varepsilon}_j\dot{\hat{\varepsilon}}_j+\chi_{1j}^{-1}\sum_{j=1}^3\tilde{W}_j^T\dot{\hat{W}}_j+\tau_{1j}^{-1}\sum_{j=1}^3\tilde{D}_j\dot{\hat{D}}_j \\
& \le s_2\{\tfrac{|\mu_{11}|\cdot|\mu_{21}|}{|h_a^2-\mu_{11}^2|}+\tfrac{|\mu_{12}|\cdot|\mu_{22}|}{|h_b^2-\mu_{12}^2|}+\tfrac{|\mu_{13}|\cdot|\mu_{23}|}{|h_c^2-\mu_{13}^2|}+\|\mu_2\|(\|\Delta\|+\|\hat{D}\|)\} \\
& +\alpha^2\rho\dot{\rho}+\gamma_{1j}^{-1}\textstyle\sum_{j=1}^3\tilde{\vartheta}_j\dot{\hat{\vartheta}}_j+\delta_{1j}^{-1}\sum_{j=1}^3\tilde{\varepsilon}_j\dot{\hat{\varepsilon}}_j+\chi_{1j}^{-1}\sum_{j=1}^3\tilde{W}_j^T\dot{\hat{W}}_j+\tau_{1j}^{-1}\sum_{j=1}^3\tilde{D}_j\dot{\hat{D}}_j\} \\
& \le -s_2L
\end{aligned}
\tag{20}
$$

Similarly, conclusion (20) ensures that the state z will reach the sliding surface s_2 within finite time. When the extended state vector z lies within the finite approximation domain of the RBFNN, i.e., when the conditions $\|\mu_2\| \le \|z\| \le \alpha|\rho|$ are satisfied, the disturbance observer, controller, and adaptive laws are designed in the following form. The disturbance observer is designed to:

$$
\hat{D} = P(t) + h(\mu_2)\mu_2, \dot{P}(t) = h(\mu_2)[u - \hat{D} - \Phi(\frac{X}{\rho})] \tag{21}
$$

Controller is designed as

$$
u = [\frac{\mu_{11}\mu_{21}}{h_a^2-\mu_{11}^2}, \frac{\mu_{12}\mu_{22}}{h_b^2-\mu_{12}^2}, \frac{\mu_{13}\mu_{23}}{h_c^2-\mu_{13}^2}]^T + \Phi(\frac{X}{\rho}) + \bar{k}\mu_2 - \hat{D} \tag{22}
$$

where $\bar{k} = \mathrm{diag}\{\bar{k}_1, \bar{k}_2, \bar{k}_3\}$, $\bar{k}_j$ is the positive real number. And the adaptive law is designed as follows:

$$
\dot{\rho} = -\zeta\rho - \sigma\alpha^2\sum_{j=1}^{3}(\hat{\vartheta}_j|\rho-1| + \alpha\hat{\varepsilon}_j)\mathrm{sign}(\rho) \tag{23}
$$

$$
\dot{\hat{\vartheta}}_j = -\tau_j\hat{\vartheta}_j + \gamma_{2j}\alpha^2|\rho|\cdot|\rho-1| \tag{24}
$$

$$
\dot{\hat{\varepsilon}}_j = -\lambda_j\hat{\varepsilon}_j + \delta_{2j}\alpha|\rho| \tag{25}
$$

$$
\dot{\hat{W}}_j = -\chi_{2j}[\Phi(\frac{X}{\rho})\mu_2^T + \eta_j\hat{W}_j] \tag{26}
$$

where $\mathrm{sign}(\rho) = \begin{cases} 1, & \rho > 0 \\ -1, & \rho \le 0 \end{cases}$, parameters $\varsigma, \sigma, \tau_j, \lambda_j, \eta_j, \gamma_{2j}, \delta_{2j}, \chi_{2j}$ are some positive constants that are given by the user. Based on (21)–(26), the following Theorem 2 is obtained. □

Theorem 2. *Under the conditions of Assumptions 1 and 2, for system* (1)*, with the disturbance observer* (21)*, controller* (22)*, and adaptive laws* (23)–(26)*, the output of system* (1) *can track the desired signal, and all signals in the closed-loop system are uniformly ultimately bounded. Proof: When* $\|\mu_2\| \le \|z\| \le \alpha|\rho|$ *is true, take*

$$
V_3 = V_2 + \frac{1}{2\sigma}\rho^2 + \sum_{j=1}^{3}\frac{1}{2\gamma_{2j}}\tilde{\vartheta}_j^2 + \sum_{j=1}^{3}\frac{1}{2\delta_{2j}}\tilde{\varepsilon}_j^2 + \sum_{j=1}^{3}\frac{1}{2\chi_{2j}}\tilde{W}_j^T\tilde{W}_j + \sum_{j=1}^{3}\frac{1}{2\tau_{2j}}\tilde{D}_j^2 \tag{27}
$$

The differential result is obtained as:

$$\begin{aligned}\dot{V}_3 &= \dot{V}_2 + \sigma^{-1}\rho\dot{\rho} + \textstyle\sum_{j=1}^{3}\frac{1}{\gamma_{2i}}\tilde{\vartheta}_j\dot{\hat{\vartheta}}_j + \sum_{j=1}^{3}\frac{1}{\delta_{2i}}\tilde{\varepsilon}_j\dot{\hat{\varepsilon}}_j + \sum_{j=1}^{3}\frac{1}{\chi_{2i}}\tilde{W}_j^T\dot{\hat{W}}_j + \sum_{j=1}^{3}\frac{1}{\tau_i}\tilde{D}_j\dot{\hat{D}}_j \\ &= -\frac{r_1\mu_{11}^2}{h_a^2-\mu_{11}^2} - \frac{r_2\mu_{12}^2}{h_b^2-\mu_{12}^2} - \frac{r_3\mu_{13}^2}{h_c^2-\mu_{13}^2} + \frac{\mu_{11}\mu_{21}}{h_a^2-\mu_{11}^2} + \frac{\mu_{12}\mu_{22}}{h_b^2-\mu_{12}^2} + \frac{\mu_{13}\mu_{23}}{h_c^2-\mu_{13}^2} - \mu_2^T(\Delta - \hat{D}) \\ &+\mu_2^T[\Delta - u - \hat{D}] + \textstyle\sum_{j=1}^{3}\gamma_{2j}^{-1}\tilde{\vartheta}_j\dot{\hat{\vartheta}}_j + \sum_{j=1}^{3}\delta_{2j}^{-1}\tilde{\varepsilon}_j\dot{\hat{\varepsilon}}_j + \sum_{j=1}^{3}\chi_{2j}^{-1}\tilde{W}_j^T\dot{\hat{W}}_j + \sum_{j=1}^{3}\tau_{2j}^{-1}\tilde{D}_j\dot{\hat{D}}_j\end{aligned} \tag{28}$$

According to (22), so it has

$$\begin{aligned}&\mu_2^T[\Delta - \Phi(\tfrac{X}{\rho})] \leq \|\mu_2\|(\textstyle\sum_{j=1}^{3}\vartheta_j\alpha|\rho-1| + \varepsilon_j) \\ &\leq \alpha|\rho|(\textstyle\sum_{j=1}^{3}\vartheta_j\alpha|\rho-1| + \varepsilon_j) = \alpha^2\sum_{j=1}^{3}(\vartheta_j|\rho|\cdot|\rho-1| + \alpha|\rho|\varepsilon_j)\end{aligned} \tag{29}$$

By (29), using (28) becomes the following form:

$$\begin{aligned}\dot{V}_3 &\leq -\frac{r_1\mu_{11}^2}{h_a^2-\mu_{11}^2} - \frac{r_2\mu_{12}^2}{h_b^2-\mu_{12}^2} - \frac{r_3\mu_{13}^2}{h_c^2-\mu_{13}^2} - \bar{k}\mu_2^T\mu_2 + \alpha^2\textstyle\sum_{j=1}^{3}\vartheta_j|\rho|\cdot|\rho-1| + \alpha|\rho|\sum_{j=1}^{3}\varepsilon_j \\ &+\sigma^{-1}\rho\dot{\rho} + \textstyle\sum_{j=1}^{3}\gamma_{2j}^{-1}\tilde{\vartheta}_j\dot{\hat{\vartheta}}_j + \sum_{j=1}^{3}\delta_{2j}^{-1}\tilde{\varepsilon}_j\dot{\hat{\varepsilon}}_j + \sum_{j=1}^{3}\chi_{2j}^{-1}\tilde{W}_j^T\dot{\hat{W}}_j + \sum_{j=1}^{3}\tau_{2j}^{-1}\tilde{D}_j\dot{\hat{D}}_j\end{aligned} \tag{30}$$

Then, the following results can be obtained by the adaptive laws (23)–(26):

$$\begin{aligned}\dot{V}_3 &\leq -\frac{r_1\mu_{11}^2}{h_a^2-\mu_{11}^2} - \frac{r_2\mu_{12}^2}{h_b^2-\mu_{12}^2} - \frac{r_3\mu_{13}^2}{h_c^2-\mu_{13}^2} - \bar{k}\mu_2^T\mu_2 + \alpha^2\textstyle\sum_{j=1}^{3}\hat{\vartheta}_j|\rho|\cdot|\rho-1| + \alpha\sum_{j=1}^{3}|\rho|\hat{\varepsilon}_j + \sigma^{-1}\rho\dot{\rho} \\ &+\textstyle\sum_{j=1}^{3}\gamma_{2j}^{-1}\tilde{\vartheta}_j(\dot{\hat{\vartheta}}_j - \gamma_{2j}\alpha^2|\rho|\cdot|\rho-1|) + \sum_{j=1}^{3}\delta_{2j}^{-1}\tilde{\varepsilon}_j(\dot{\hat{\varepsilon}}_j - \alpha\delta_{2j}|\rho|) \\ &+\textstyle\sum_{j=1}^{3}\chi_{2j}^{-1}\tilde{W}_j[-\chi_{2j}\Phi(\frac{X}{\rho})\mu_2^T + \eta\dot{\hat{W}}_j] + \sum_{j=1}^{3}\tau_{2j}^{-1}\tilde{D}_j\dot{\tilde{D}}_j \\ &\leq -\frac{r\mu_1^2}{h_a^2-\mu_1^2} - \bar{k}\mu_2^T\mu_2 - \frac{\zeta}{\sigma}\rho^2 - \textstyle\sum_{j=1}^{3}\frac{\tau_j}{\gamma_{2j}}\tilde{\vartheta}_j\hat{\vartheta}_j - \sum_{j=1}^{3}\frac{\lambda_j}{\delta_{2j}}\tilde{\varepsilon}_j\hat{\varepsilon}_j - \sum_{j=1}^{3}\frac{\eta_j}{\chi_{2j}}\tilde{W}_j^T\hat{W}_j \\ &+\textstyle\sum_{j=1}^{3}\frac{1}{2\tau_{2j}}\tilde{D}_j^2 + \sum_{j=1}^{3}\frac{1}{2\tau_{2j}}\dot{\tilde{D}}_j^2\end{aligned} \tag{31}$$

In Formula (31), $\mu_1^2 - \text{diag}\{\mu_{11}^2, \mu_{12}^2, \mu_{13}^2\}$. Because the following inequalities hold:

$$-\frac{\tau_j}{\gamma_{2j}}\tilde{\vartheta}_j\hat{\vartheta}_j = -\frac{\tau_j}{\gamma_{2j}}\tilde{\vartheta}_j^2 - \frac{\tau_j}{\gamma_{2j}}\tilde{\vartheta}_j\vartheta_j \leq -\frac{\tau_j}{2\gamma_{2j}}\tilde{\vartheta}_j^2 + \frac{\tau_j}{2\gamma_{2j}}\vartheta_j^2 \tag{32}$$

$$-\frac{\lambda_j}{\delta_{2j}}\tilde{\varepsilon}_j\hat{\varepsilon}_j = -\frac{\lambda_j}{\delta_{2j}}\tilde{\varepsilon}_j^2 - \frac{\lambda_j}{\delta_{2j}}\tilde{\varepsilon}_j\varepsilon_j \leq -\frac{\lambda_j}{2\delta_{2j}}\tilde{\varepsilon}_j^2 + \frac{\lambda_j}{2\delta_{2j}}\varepsilon_j^2 \tag{33}$$

$$-\frac{\eta_j}{\chi_{2j}}\tilde{W}_j^T\hat{W}_j = -\frac{\eta_j}{\chi_{2j}}\tilde{W}_j^T\tilde{W}_j - \frac{\eta_j}{\chi_{2j}}\tilde{W}_j^T\hat{W}_j \leq -\frac{\eta_j}{2\chi_{2j}}\tilde{W}_j^T\tilde{W}_j + \frac{\eta_j}{2\chi_{2j}}\tilde{W}_j^T\hat{W}_j. \tag{34}$$

With inequalities (28)–(30), it can be obtained as:

$$\begin{aligned}\dot{V}_3 &\leq -\frac{r\mu_1^2}{h_a^2-\mu_1^2} - \bar{k}\mu_2^T\mu_2 - \frac{\zeta}{\sigma}\rho^2 + \textstyle\sum_{j=1}^{3}\frac{\tau_j}{\gamma_{2j}}\vartheta_j^2 + \sum_{j=1}^{3}\frac{\lambda_j}{\delta_{2j}}\varepsilon_j^2 + \sum_{j=1}^{3}\frac{\eta_j}{\chi_{2j}}\tilde{W}_j^T\tilde{W}_j + \sum_{j=1}^{3}\frac{\eta_j}{\chi_{2j}}\tilde{W}_j^T\hat{W}_j \\ &+\textstyle\sum_{j=1}^{3}\frac{1}{2\tau_{2j}}\tilde{D}_j^2 + \sum_{j=1}^{3}\frac{1}{2\tau_{2j}}\dot{\tilde{D}}_j^2\end{aligned} \tag{35}$$

Define

$$\begin{aligned}&\omega = \min\{r_1, r_2, r_3, \bar{k}_1, \bar{k}_2, \bar{k}_3, \zeta\} \\ &\sigma = \textstyle\sum_{j=1}^{3}\frac{\tau_j}{\gamma_{2j}}\vartheta_j^2 + \sum_{j=1}^{3}\frac{\lambda_j}{\delta_{2j}}\varepsilon_j^2 + \sum_{j=1}^{3}\frac{\eta_j}{\chi_{2j}}\tilde{W}_j^T\hat{W}_j + \sum_{j=1}^{3}\frac{1}{2\tau_{2j}}\tilde{D}_j^2 + \sum_{j=1}^{3}\frac{1}{2\tau_{2j}}\dot{\tilde{D}}_j^2\end{aligned} \tag{36}$$

Then (31) is equivalent to:

$$\dot{V}_3 \leq -\omega V_3 + \sigma \tag{37}$$

Multiply $e^{\omega t}$ on the both sides of the Equation (32) and integrate in the interval $[0, t]$ to get:

$$0 \leq V_3(t) \leq [V_3(0) - \frac{\sigma}{\omega}]e^{-\omega t} + \frac{\sigma}{\omega} \tag{38}$$

From (33), it can be concluded that $\bar{V}(t)$ is bounded. Therefore, the tracking error μ_1, signals μ_2, observer error $\tilde{D}$, adaptive parameters ρ, parameter estimates $\hat{\theta}$, $\hat{\varepsilon}$ and the estimated RBFNN weight vector, $\tilde{W}_j$ can converge to a compact set, $\Omega = [V_3(0) - \frac{\sigma}{\varpi}]e^{-\omega t} + \frac{\sigma}{\varpi}$ can converge to a compact set.

Remark 2. *To obtain higher approximation accuracy, it is typical to utilize a large number of neurons in RBFNN control design. Nevertheless, the approximation accuracy* (25) *can be fine-tuned by the parameter* (23)*, which removes the requirement for a large number of neurons in RBFNNs.*

3. Simulation Verification of RBFNN-Based Adaptive Robust Control

Consider the desired tracking trajectory defined by the following continuous functions along the XYZ-axis, where all three functions are smooth and differentiable.

$$y_d = [y_{d1}, y_{d2}, y_{d3}]^T = [-0.5\cos(\pi t), 0.3 - 0.05\cos(\pi t), 0.35\sin(\pi t)\cos(0.5t)]^T$$

Three kinds of external disturbance are considered in the simulation.

The initial values in the FOG are given as Table 1.

Table 1. Initial values in the fiber optic gyro control system.

Parameter	Value	Parameter	Value	Parameter	Value
ϕ_{F1}	-0.4	ϕ_{F2}	0.22	ϕ_{F3}	-0.15
ϕ_{s1}	$-\pi$	ϕ_{s2}	π	ϕ_{s3}	-0.5π
k_{11}	3	k_{12}	3	k_{13}	3
k_{21}	10	k_{22}	10	k_{23}	10
α	100	L	10	h_{a1}	5
h_{b1}	-2	h_{b2}	-3	h_{c1}	-1
τ_1	100	τ_2	200	τ_3	300
$\rho_1(0)$	0.6	$\rho_2(0)$	0.9	$\rho_3(0)$	0.3
γ_{21}	0.001	γ_{22}	0.002	γ_{21}	0.003
r_1	3	r_2	3	r_3	3
δ_{11}	0.002	δ_{22}	0.004	δ_{23}	0.006
$\bar{k}_1$	100	$\bar{k}_2$	200	$\bar{k}_3$	300
λ_1	250	λ_2	350	λ_3	450
ξ	20	σ	10	$\rho(0)$	0.1
$\vartheta_1(0)$	0.6	$\vartheta_2(0)$	0.9	$\vartheta_3(0)$	0.3
$\varepsilon_1(0)$	0.5	$\varepsilon_2(0)$	0.7	$\varepsilon_3(0)$	0.4

Simulation 1: When the external disturbance is $D = [D_1, D_2, D_3]^T = [2\cos(3t), 2\sin(5t), 2\sin(5t)]^T$, where each function are is smooth and differentiable.

In Figure 2, the output of the FOG control system (1) effectively tracks the desired reference signal, and the time response of the tracking error is shown in Figure 3. The simulation results indicate that the tracking error converges. The actual disturbance of the system and the simulation results of the disturbance observer are shown in Figure 4, while the time response of the observation error is presented in Figure 5. These results demonstrate that the disturbance observer designed in this study achieves high-precision observation. The corresponding controller and the parameter adaptation laws are shown in Figures 6–9. Figure 6 illustrates that the overall time-varying trend of the controller remains stable. Figure 7 shows the time response of the virtual controller. Figure 8 depicts the time response of the adaptive parameters within the controller. The variation of the Lipschitz adaptive parameters in Figure 8 indicates the existence of a series of suitable small constants. In Figure 9, the nonzero parameters within the controller can automatically adjust online and reach appropriate values, thereby satisfying the optimal approximation performance of the RBFNN. The approximation accuracy of the RBFNN can automatically adjust and asymptotically approach zero. These simulation results demonstrate that the RBFNN adaptive controller designed in this study effectively achieves the desired system tracking and high-precision tracking and observation of unknown external disturbances.

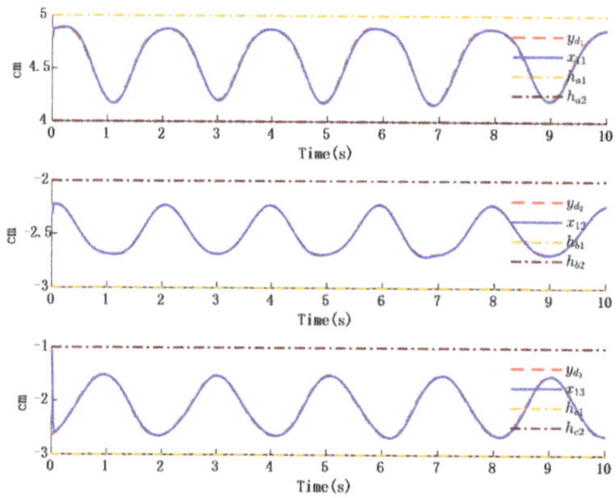

Figure 3. Time response of position tracking error.

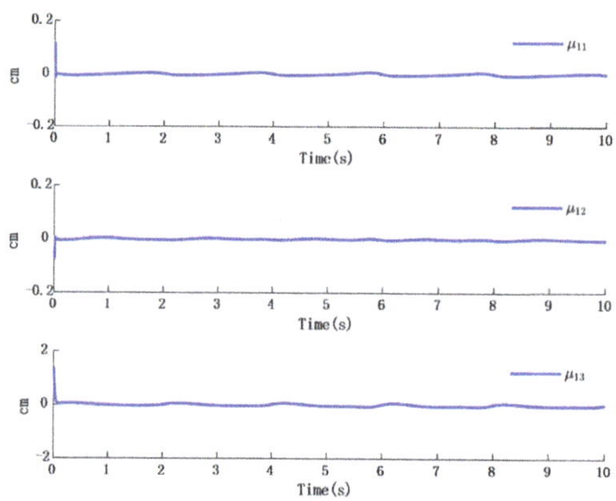

Figure 4. Time response of position tracking error.

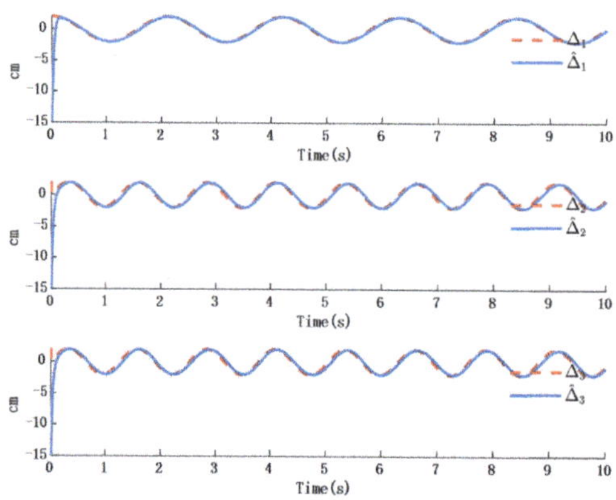

Figure 5. Time response of actual disturbance and disturbance observer error.

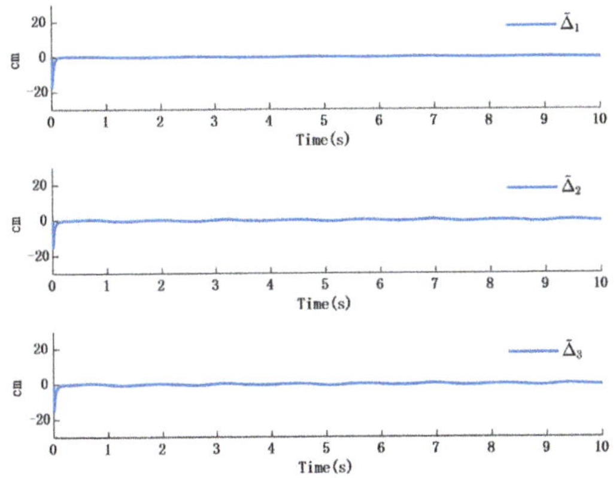

Figure 6. Time response of actual disturbance and disturbance observer error.

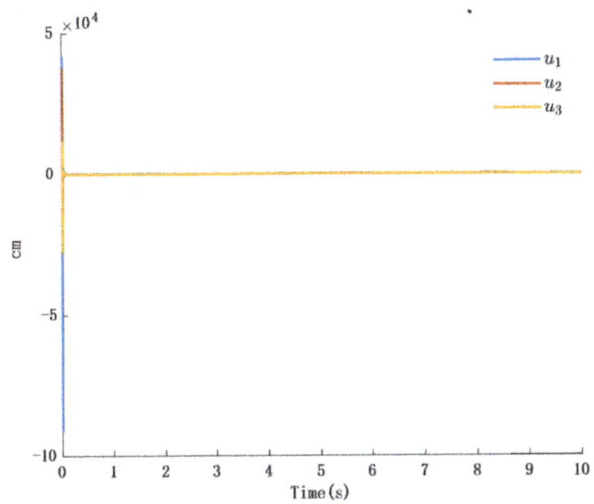

Figure 7. Time response of actual disturbance and disturbance observer error.

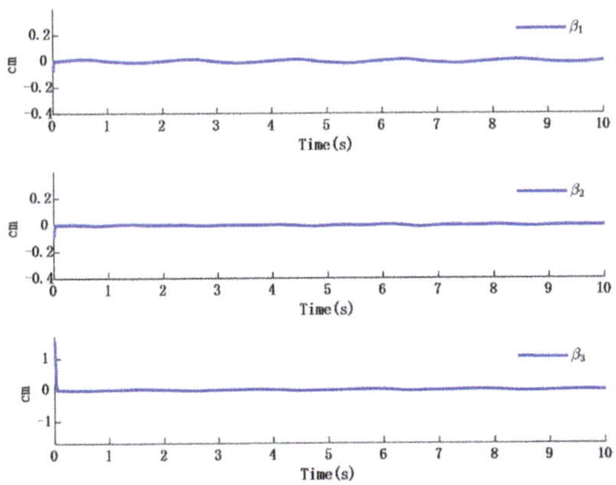

Figure 8. Time response of the controller β.

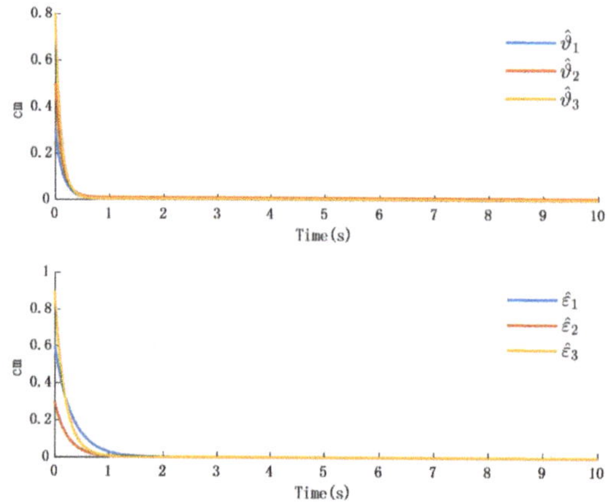

Figure 9. Time response of adaptive parameters.

Simulation 2: When the external disturbance is $D = [D_1, D_2, D_3]^T$ a step response as shown in Table 2, where the three external disturbances along the XYZ-axis are modeled as non-differentiable piecewise functions, commonly named as square wave interference, the simulation results are presented in Figures 10 and 11:

Table 2. Disturbance values produced in time 0 to 10.

Time (Second)	$0 \leq t \leq 2$	$2 < t \leq 4$	$4 < t \leq 6$	$6 < t \leq 8$	$8 < t \leq 10$
D_1	−20	20	−20	20	−20
D_2	−20	20	−20	20	−20
D_3	−20	20	−20	20	−20

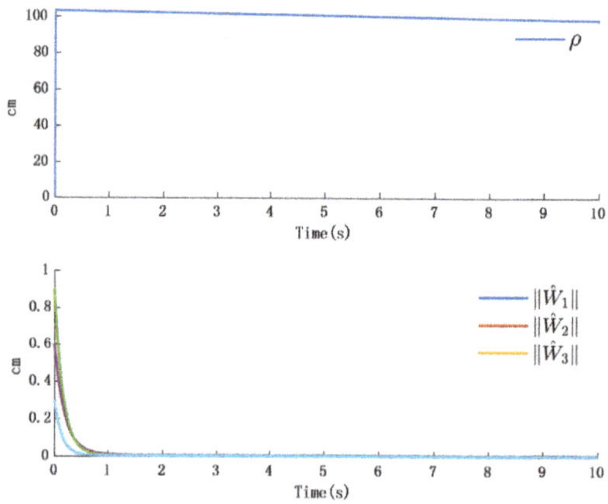

Figure 10. Time response of adaptation in the controller.

The simulation results in Figure 10 illustrate the comparison between the designed disturbance observer and the actual disturbance under step responses occurring at different time intervals. It can be known from Figure 9 that the time response of the observer designed in this paper is almost the same as that in the actual interference situation. Since the time response in the actual situation is also affected by other factors, such as environmental temperature, etc., a certain delay will occur in the system when converting between high and low levels. However, in the stable state, the observer designed in this paper is consistent with the time response characteristics of the actual situation. Therefore, it will not have an impact on the performance of the system. Figure 11 presents the time response of the error between the actual step disturbance and the observed step disturbance. It is obvious that the error between the actual step response and the observed step response time is consistent. It can therefore be concluded that the designed disturbance observer achieves effective observation of external disturbances even when the disturbances take the form of step signals.

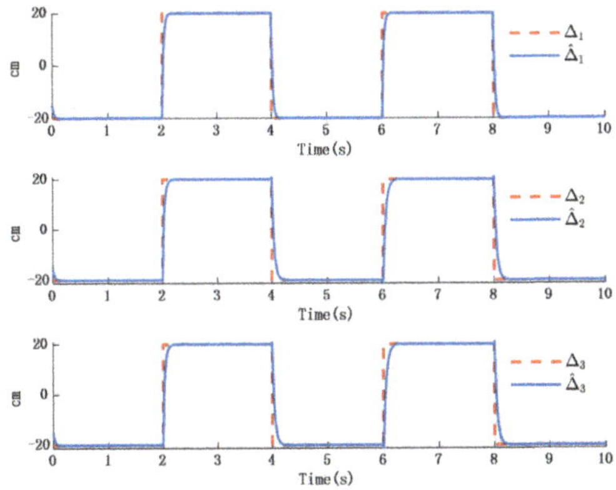

Figure 11. Step time response and time response of the observer.

Simulation 3: When the external disturbance is $D = [D_1, D_2, D_3]^T$. In this case, each function includes a ramp signal as described below, where the magnitude of the ramp signal increases linearly over time.

$$D_j = \begin{cases} 3t, & 0 \leq t < 2 \\ 3t - 2, & 2 \leq t < 4 \\ 3t - 4, & 4 \leq t < 6 \\ 3t - 6, & 6 \leq t < 8 \\ 3t - 8, & 8 \leq t < 10 \end{cases}$$

Figures 12 and 13 shows the simulation results; Figure 14 shows error time response of slope and observer:

In Figure 12, the time response of the observer aligns with the external disturbance when the disturbance takes the form of a ramp signal. Figure 13 shows that the time response of the error between the ramp disturbance and the observer converges to zero.

From the simulation results of three different types of external disturbances, it can be concluded that the proposed design method enables the output of the FOG system (1) to effectively track the desired reference signal. Moreover, the external disturbances can be effectively observed through the designed observer. The unknown dynamic model within the system can be effectively approximated using the designed RBFNN, with the approximation accuracy being adaptively regulated through the controller parameters, significantly improving the approximation precision. Furthermore, all signals in the closed-loop system are ensured to achieve uniform ultimate boundedness.

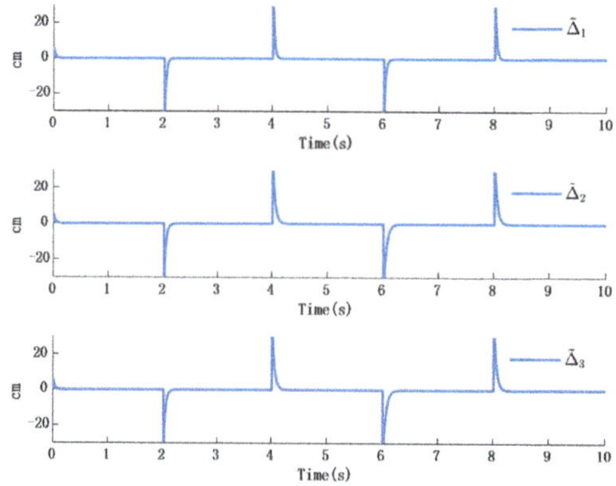

Figure 12. Time response of the error between step response and observer.

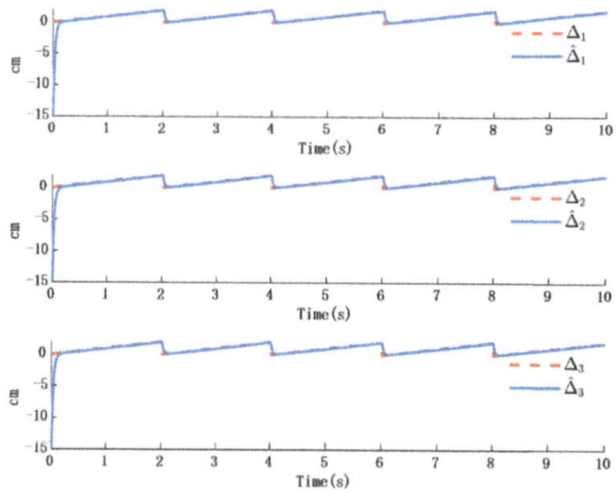

Figure 13. Time response of slope and observer.

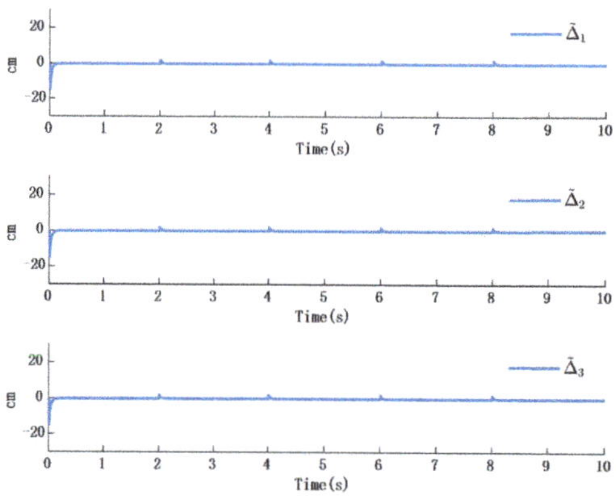

Figure 14. Error time response of slope and observer.

Remark 3. *In the first simulation, a smooth and differentiable desired trajectory along the XYZ-axis. The results show that the system can achieve accurate tracking performance, and the disturbance observer error converges to zero. In the second simulation, a step-like external disturbance, which is non-differentiable at the inflection point due to its piecewise nature. This represents a different class of disturbance compared to the first case. Finally, in the third simulation, a piecewise* *increasing function as the external disturbance, which presents a more complex and challenging scenario than the previous two cases. Despite these varying levels of complexity such as increasing, the simulation results consistently demonstrate good tracking performance, and the disturbance observer effectively estimates and compensates for the unknown external disturbances.*

Remark 4. *This study models the dynamical equation of the Digital closed-loop FOG, differing from other models that linearize the nonlinear terms. In this work, the RBFNNs to approximate the unknown nonlinear terms. Therefore, comparisons with other methods for same class of Digital closed-loop FOG is the main research work in future. because of the unique model of the digital closed-loop FOG.*

Remark 5. *The main focus of this work is on the design of RBFNN adaptive control theory for FOG. Experimental testing is a key task due to more complex factors such as hardware and others, it is our main work in future research.*

4. Conclusions

A new RBFNN-based adaptive robust control technique is proposed for high-precision angular position and velocity tracking of FOG under large dynamic unknown external disturbances. The designed controller not only relaxes the limitations imposed by the finite domain of the RBFNN but also enables online adjustment based on the desired control accuracy. Compared to the simulation results under the different unknown external disturbance, the RBFNN-based adaptive robust control technique demonstrates the ability to achieve high-precision tracking and effective disturbance observation in the presence of unknown external disturbances. Indeed, there are numerous complex factors that can influence the system, such as hardware implementation, unbounded disturbances, and integration with multi-sensor systems. These various elements pose challenges in practical applications. The research conducted can provide significant importance for enhancing the performance of high-dynamic FOG-based strapdown inertial navigation systems.

Author Contributions: Conceptualization, S.L., B.L. and J.M.; methodology, S.L. and S.L.; software, S.L.; validation, S.L. and B.L.; formal analysis, S.L.; investigation, S.L.; resources, J.M.; data curation, S.L.; writing—original draft preparation, S.L.; writing—review and editing, J.M. and X.D.; visualization, S.L.; supervision, H.L.; project administration, H.L. All authors have read and agreed to the published version of the manuscript.

Funding: National Natural Science Foundation of China (NSFC) (62163006, 61605153).

Institutional Review Board Statement: Not applicable.

Informed Consent Statement: Not applicable.

Data Availability Statement: The original contributions presented in this study are included in the article. Further inquiries can be directed to the corresponding author.

Conflicts of Interest: The authors declare no conflicts of interest.

References

1. Udd, E. An overview of the development of fiber gyros. *Opt. Waveguide Laser Sens.* **2020**, *11405*, 1140502.
2. Wang, Q.; Yang, C.; Wang, X.; Wang, Z. All-digital signal-processing open-loop fiber-optic gyroscope with enlarged dynamic range. *Opt. Lett.* **2013**, *38*, 5422–5425. [CrossRef] [PubMed]
3. Jin, J.; Ren, C.; Teng, F.; Zhang, S. Method of suppression of impulse interferences in digital closed loop fiber optic gyro detected signal. *Acta Astronaut.* **2017**, *130*, 162–166. [CrossRef]
4. Zhao, S.; Zhou, Y.; Shu, X. Analysis of fiber optic gyroscope dynamic error based on CEEMDAN. *Opt. Fiber Technol.* **2022**, *69*, 102835. [CrossRef]

5. Mou, J.; Huang, T.; Shu, X. Error analysis and comparison of the fiber optic gyroscope scale factor obtained by angular velocity method and angular increment method. *Mapan-J. Metrol. Soc. India* **2020**, *35*, 407–419. [CrossRef]
6. Zheng, Y.; Zhang, C.; Li, L.; Song, L. Loop gain stabilizing with an all-digital automatic-gain-control method for high-precision fiber-optic gyroscope. *Appl. Opt.* **2016**, *55*, 4589–4595. [CrossRef]
7. Pogorelaya, D.A.; Smolovik, M.A.; Volkovskiy, S.A.; Mikheev, M. Adjustment of PID controller in fiber-optic gyro feedback loop. *Gyroscopy Navig.* **2017**, *8*, 235–239. [CrossRef]
8. Li, Q.; Ben, Y.; Sun, F. Strapdown fiber optic gyrocompass using adaptive network-based fuzzy inference system. *Opt. Eng.* **2014**, *53*, 014103. [CrossRef]
9. Pan, Y.; Du, P.; Xue, H.; Lam, H.-K. Singularity-free fixed-time fuzzy control for robotic systems with user-defined performance. *IEEE Trans. Fuzzy Syst.* **2020**, *29*, 2388–2398. [CrossRef]
10. Fei, J.; Xin, M. An adaptive fuzzy sliding mode controller for MEMS triaxial gyroscope with angular velocity estimation. *Nonlinear Dyn.* **2012**, *70*, 97–109. [CrossRef]
11. Şerbetçi, H.; Navruz, İ. Kapalı Döngü Fiberoptik Jiroskop Sistemleri için Yerçekimi Arama Algoritmasına Dayalı PID Kontrolcü Tasarımı. *Int. J. Eng. Res. Dev.* **2019**, *11*, 695–704.
12. Shao, X.; Shi, Y. Neural adaptive control for MEMS gyroscope with full-state constraints and quantized input. *IEEE Trans. Ind. Inform.* **2020**, *16*, 6444–6454. [CrossRef]
13. Li, Z.; Yang, C.; Tang, Y. Decentralised adaptive fuzzy control of coordinated multiple mobile manipulators interacting with non-rigid environments. *IET Control Theory Appl.* **2013**, *7*, 397–410. [CrossRef]
14. Ding, J.; Zhang, J.; Huang, W.; Chen, S. Laser gyro temperature compensation using modified RBFNN. *Sensors* **2014**, *14*, 18711–18727. [CrossRef]
15. Zirkohi, M.M. Adaptive interval type-2 fuzzy recurrent RBFNN control design using ellipsoidal membership functions with application to MEMS gyroscope. *ISA Trans.* **2022**, *119*, 25–40. [CrossRef] [PubMed]
16. Fei, J.; Ding, H.; Hou, S.; Wang, S.; Xin, M. Robust adaptive neural sliding mode approach for tracking control of a MEMS triaxial gyroscope. *Int. J. Adv. Robot. Syst.* **2012**, *9*, 20. [CrossRef]
17. Xin, M.; Fei, J. Adaptive vibration control for MEMS vibratory gyroscope using backstepping sliding mode control. *J. Vib. Control* **2015**, *21*, 808–817. [CrossRef]
18. Fei, J.; Batur, C. A novel adaptive sliding mode control with application to MEMS gyroscope. *ISA Trans.* **2009**, *48*, 73–78. [CrossRef]
19. Zirkohi, M.M. Adaptive backstepping control design for MEMS gyroscope based on function approximation techniques with input saturation and output constraints. *Comput. Electr. Eng.* **2022**, *97*, 107547. [CrossRef]
20. Liang, H.; Liu, G.; Huang, T.; Lam, H.-K.; Wang, B. Cooperative fault-tolerant control for networks of stochastic nonlinear systems with nondifferential saturation nonlinearity. *IEEE Trans. Syst. Man Cybern. Syst.* **2020**, *52*, 1362–1372. [CrossRef]
21. Kong, L.; He, W.; Dong, Y.; Cheng, L.;Yang, C.; Li, Z. Asymmetric bounded neural control for an uncertain robot by state feedback and output feedback. *IEEE Trans. Syst. Man Cybern. Syst.* **2019**, *51*, 1735–1746. [CrossRef]
22. Yang, C.; Huang, D.; He, W.; Cheng, L. Neural control of robot manipulators with trajectory tracking constraints and input saturation. *IEEE Trans. Neural Netw. Learn. Syst.* **2020**, *32*, 4231–4242. [CrossRef] [PubMed]

biomimetics MDPI

Article

YOLO-HVS: Infrared Small Target Detection Inspired by the Human Visual System

Xiaoge Wang [1], Yunlong Sheng [1,*], Qun Hao [2], Haiyuan Hou [1] and Suzhen Nie [1]

1 School of Mechanical Engineering, Shandong University of Technology, Zibo 255000, China; wangxg124@163.com (X.W.); houhaiyuan11@163.com (H.H.); szhen0203@163.com (S.N.)
2 Changchun University of Science and Technology, Changchun 130022, China; qhao@bit.edu.cn
* Correspondence: shengyunlong@sdut.edu.cn

Abstract

To address challenges of background interference and limited multi-scale feature extraction in infrared small target detection, this paper proposes a YOLO-HVS detection algorithm inspired by the human visual system. Based on YOLOv8, we design a multi-scale spatially enhanced attention module (MultiSEAM) using multi-branch depth-separable convolution to suppress background noise and enhance occluded targets, integrating local details and global context. Meanwhile, the C2f_DWR (dilation-wise residual) module with regional-semantic dual residual structure is designed to significantly improve the efficiency of capturing multi-scale contextual information by expanding convolution and two-step feature extraction mechanism. We construct the DroneRoadVehicles dataset containing 1028 infrared images captured at 70–300 m, covering complex occlusion and multi-scale targets. Experiments show that YOLO-HVS achieves mAP50 of 83.4% and 97.8% on the public dataset DroneVehicle and the self-built dataset, respectively, which is an improvement of 1.1% and 0.7% over the baseline YOLOv8, and the number of model parameters only increases by 2.3 M, and the increase of GFLOPs is controlled at 0.1 G. The experimental results demonstrate that the proposed approach exhibits enhanced robustness in detecting targets under severe occlusion and low SNR conditions, while enabling efficient real-time infrared small target detection.

Keywords: infrared small target detection; human visual system; YOLOv8; attention mechanism

1. Introduction

Infrared small target detection (IRSTD) utilizes thermal radiation discrepancies between targets and their backgrounds in infrared imaging to identify and track objects of interest. Due to its strong penetration, good concealment, strong anti-jamming ability, and independence of light conditions, it has found extensive applications across multiple domains, including military operations, surveillance systems, early warning mechanisms, precision targeting, and maritime reconnaissance [1,2]. Compared with ordinary visible targets, infrared weak target imaging distance is long, resulting in the target pixels accounting for a small proportion of the pixels of the entire image and the lack of obvious texture, shape, and structure features [3,4]. Secondly, the energy of infrared radiation decays significantly with the increase of distance, which makes the infrared weak targets easily submerged in the background clutter and noise. The detection of infrared small targets remains technically challenging in contemporary research due to the extremely low pixel values of the small targets, their extremely low contrast, and the extremely close proximity of the small targets. In addition, real-time target output is required in many application scenarios, so detection algorithms with fast detection speed and a low number of model parameters are more popular [5].

With the development of deep convolutional neural networks, the performance of infrared small target detectors has been significantly enhanced. Many high-performance infrared small target detection algorithms based on deep learning have been put forward [6,7]. In general, these algorithms can be categorized into two branches. One branch of deep learning-based infrared small target detection algorithms employs a two-stage neural network as both a feature extractor and classifier, enabling the detection of infrared small targets from coarse to fine scales. Although the two-stage detector has achieved great success, it also suffers from disadvantages such as difficulty in training and slow detection speed [8]. Another branch is single-stage detection algorithms, which improve the detection speed by performing classification and regression directly on the feature map. Representatives of single-stage detection algorithms, such as YOLO [9,10], reduce the computational complexity by simplifying the detection process, thus significantly increasing the detection speed while maintaining higher accuracy [11,12].

Although deep convolutional networks significantly improve IR small target detection, detecting IR small targets with large variations in scale, occlusion, appearance, and illumination in real-world scenes is still a great challenge. Recent studies in infrared small target detection have incorporated functional characteristics inspired by the human visual system (HVS) to improve detection performance, encompassing size adaptation dynamics, contrast perception mechanisms, and attentional modulation processes. These introductions notably improve key detection capabilities, ranging from identification accuracy to computational efficiency [13–15]. As exemplified by the attention mechanism, its design originates from the human visual system's selective attention in information processing: humans tend to prioritize task-relevant information when dealing with abundant data [16]. In deep learning, the attention mechanism enhances model performance by dynamically allocating higher weights to salient input features while suppressing less relevant ones, moving beyond uniform treatment of all inputs. Ref. [17] repaired the Null Convolutional Block Attention Module (DILATED CBAM) to YOLOv4's backbone network, CSPDarknet53, to augment the small targets and make key features distinguishable. Ref. [18] proposed a dilated convolution-based weighted feature pyramid network (DWFPN). The DWFPN assigns fusion weights to multi-level features via an attention mechanism, facilitating high-quality information interaction while preventing the loss of small target features and critical information during network deepening.

Although a range of techniques have been utilized for studying infrared small target detection from different angles, the existing research still faces inherent constraints that need to be addressed [19]. On the one hand, the interference of background information is an important issue during IR small target detection. In operational scenarios, targets are commonly overshadowed by environmental noise—including vegetation, shadows, and other surface features [20]. Background interference significantly degrades detection performance by introducing false positives (misclassifying negative samples as targets), thereby compromising model accuracy. Furthermore, existing advanced methods struggle with multi-scale target extraction due to the substantial size variations typical of infrared small targets, leading to unreliable detection across different scales [21,22].

Aiming to resolve the identified limitations, this research proposes a target detection algorithm termed YOLO-HVS based on YOLOv8 and HVS. Firstly, for the purpose of reducing the impact of non-relevant information obscuring small IR targets in the detection process, the MultiSEAM (Multi-scaled Spatially Enhanced Attention Module) attention mechanism is introduced. Secondly, in order to expand the sensing field to extract rich multi-scale contextual information, C2f_DWR is designed to replace the C2f module in YOLOv8. This paper makes the following main contributions:

(1) This study constructs a multi-scenario infrared vehicle dataset containing occluded and multi-scale targets. We propose YOLO-HVS to enhance small vehicle detection, validated on both DroneVehicle and our DroneRoadVehicles datasets.

(2) To address IR small target occlusion, we integrate MultiSEAM into YOLOv8. This attention mechanism enhances unoccluded features through multi-branch depth-wise convolutions, capturing both local details and global context for robust detection.

(3) To enhance multi-scale context extraction, we propose the C2f_DWR module (C2f with dilation-wise residual) as a replacement for YOLOv8's original C2f. This dual-residual structure splits feature extraction into regional and semantic pathways, simplifying multi-scale mapping while improving detection efficiency across varying target scales.

2. Datasets

The two datasets used for the experiments in this paper are now described.

2.1. *Public Dataset-DroneVehicle*

The DroneVehicle dataset [23], a comprehensive collection of aerial vehicle images captured by UAVs and publicly released by Tianjin University, serves as a critical benchmark for advancing computer vision research in unmanned aerial systems. The dataset covers rich RGB and infrared image resources for vehicle detection and counting tasks and covers diverse scenes from day to night. Given that the research focus of this paper lies in the domain of infrared small targets, only the infrared images within the dataset are employed as research objects.

In the experimental design, we strictly adhere to the public dataset partitioning strategy and partition the dataset into a training set and a validation set. Specifically, the training set is composed of 17,990 images, while the validation set includes 1469 images. The dataset encompasses five primary classes: car, truck, bus, van, and freight car. These classes cover typical vehicle types in UAV aerial photography scenarios, offering abundant sample support for infrared small target detection tasks.

2.2. *Our Dataset-DroneRoadVehicles*

It is observed that the vehicle targets within the DroneVehicle dataset exhibit relatively uniform sizes, which may be due to the relatively fixed flight altitude of the UAVs during the filming process. However, in practical applications, the flight altitude of the UAV often varies depending on the mission requirements, and this single-view shooting condition may be difficult to generalize in real-world scenarios [24]. To validate the versatility and stability of the proposed approach across varied environmental conditions, a novel dataset named DroneRoadVehicles was developed. This dataset spans an altitude range of 70 m to 300 m by capturing images at distinct flight heights, thereby encompassing large, medium, and small vehicle targets with diverse size distributions. Such a design not only augments dataset diversity but also furnishes more challenging samples for infrared small target detection tasks, enabling the validation of the method's effectiveness and generalizability in complex scenarios.

The data collection device employed is the DJI MAVIC2 Enterprise Advanced, with its thermal and visible camera specifications fundamentally outlined in Table 1.

Table 1. Camera basic parameters.

Indicators	Thermal Camera	Visual Camera
Spectral Band	(8–14) μm	(0.38–0.7) μm
Resolution	640 × 512	3840 × 2160/1920 × 1080
Sensors	Uncooled VOx Microbolometer	1/2 CMOS

During infrared vehicle data acquisition, significant efforts were made to account for real-world environmental interference. The collected data span multiple diurnal cycles (daytime to nighttime) and diverse meteorological conditions (e.g., clear skies and fog), ensuring comprehensive environmental coverage. Furthermore, the dataset incorporates a wide range of practical scenarios, including but not limited to expressways, parking facilities, commercial centers, and urban crossroads. This design ensures that the dataset is diverse in terms of scenarios and environments, thus providing strong support for validating the adaptability and robustness of various types of network models under complex conditions.

The DroneRoadVehicles dataset contains 1028 aligned infrared-visible image pairs, divided into training (80%), validation (10%), and test sets (10%) following a standardized data split protocol. As illustrated in Figure 1, representative samples demonstrate multi-modal scene coverage, where visible-spectrum images (top row) are precisely registered with their thermal infrared equivalents (bottom row) across varied environments. Figure 2 presents UAV-captured images at varying flight altitudes, further highlighting the dataset's target size diversity. As illustrated in Figure 3, the dataset's primary categories encompass cars, trucks, and buses. The diversity of scenes and environments within this dataset facilitates the testing of infrared vehicle target detection and tracking tasks.

(**a**) Urban highway (sunny day) (**b**) Parking (night) (**c**) Crossroad

Figure 1. Multi-scene examples of our DroneRoadVehicles dataset.

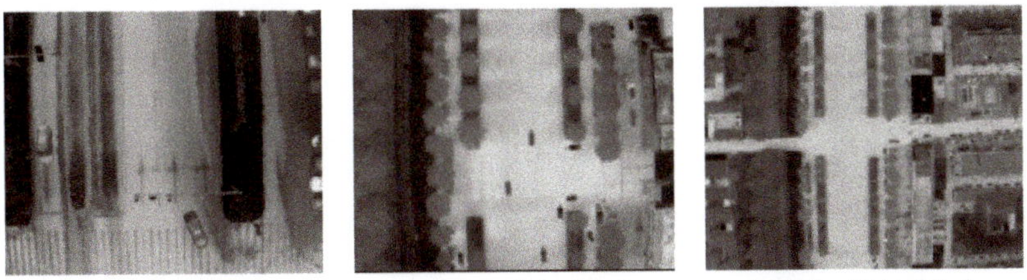

(**a**) Lower flight altitude (**b**) Medium flight altitude (**c**) Higher flight altitude

Figure 2. Comparison of drones at different flight altitudes.

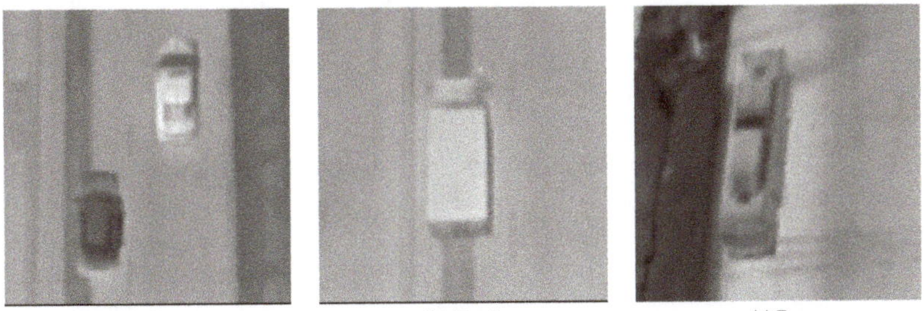

(**a**) Car (**b**) Truck (**c**) Bus

Figure 3. Target vehicle categories in the DroneRoadVehicles dataset.

3. Methods

3.1. YOLO-HVS

YOLO-HVS represents an enhanced detection architecture built upon YOLOv8, specifically tailored for infrared small object detection tasks [25]. The overall framework is visualized in Figure 4. During the optimization of the baseline YOLOv8, the following critical strategies were implemented:

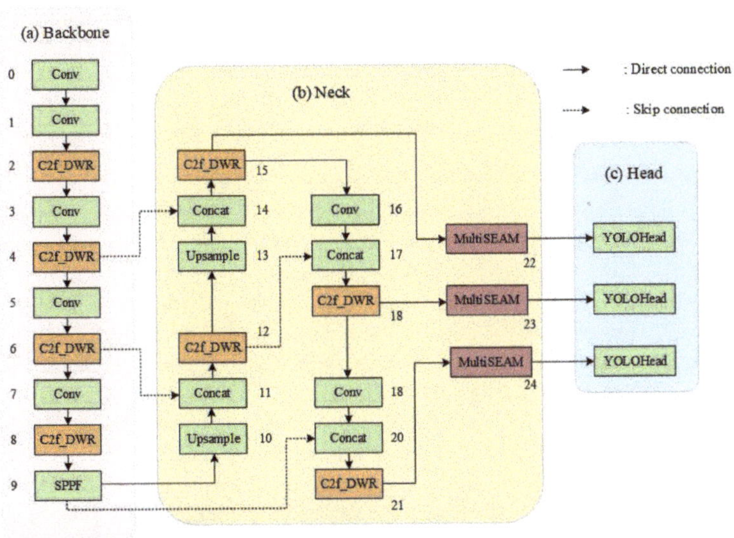

Figure 4. Overall network architecture of YOLO-HVS.

First, three MultiSEAM attention mechanisms are introduced at the end of Neck. This mechanism effectively addresses the issue of inaccurate localization arising from target feature loss due to obstruction by surrounding objects. The attention module is designed to enhance target feature learning; meanwhile, it improves feature map resolution, thereby boosting detection precision for small infrared targets.

Second, we innovatively adopt C2f_DWR to replace the original C2f in YOLOv8. This improvement effectively expands the receptive field and significantly reduces the difficulty of mapping multi-scale contextual information. The expanded receptive field enables the model to integrate contextual information across broader spatial regions, thereby improving its ability to analyze target-environment interactions. Simultaneously, easing the challenge of extracting multi-scale contextual information enables the model to handle targets across diverse scales more adeptly. This optimization significantly improves the model's capacity to detect targets across varying scales, allowing for more precise localization and identification of small infrared targets across various scales.

3.2. MultiSEAM Attention Mechanism

In target detection, it often occurs that the target is occluded by other objects, which can lead to the disappearance of features, thus leading to inaccurate localization, and at the same time, taking into account the characteristics of infrared small target detection, we innovatively introduced a multi-branching SEAM (Self-Ensembling Attention Mechanism), known as MultiSEAM [26]. As shown in Figure 5 for the schematic diagram of the MultiSEAM Attention Mechanism.

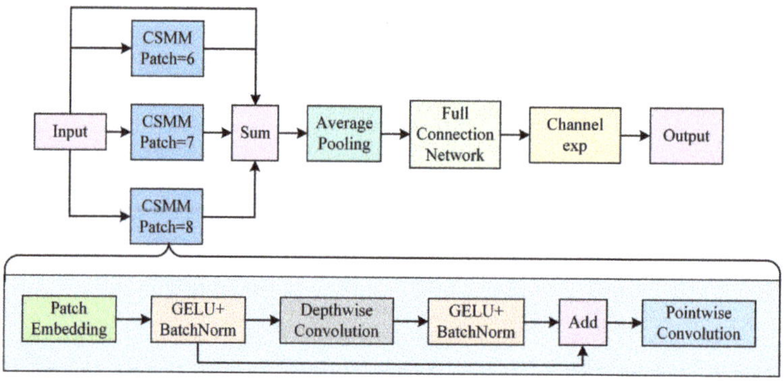

Figure 5. The MultiSEAM architecture (top) and its Channel-Spatial Mixing Module (CSMM, down).

The core design of the MultiSEAM module is shown in Figure 5, which demonstrates a multi-branch parallel architecture consisting of three CSMMs (Contextual Spatial-Channel Modeling Module) of different sizes (patch = 6/7/8) on the top side. In the feature processing flow, the input image is first segmented into multi-scale local regions (6 × 6, 7 × 7, and 8 × 8), which are mapped to a high-dimensional space by a Patch Embedding layer with shared weights to generate the initial feature representation. Each CSMM module models joint spatial-channel correlations through a two-stage operation. (1) Depthwise convolution is when the spatial features of each channel are extracted independently using depthwise convolution [27], which captures the local context information using a 3 × 3 convolution kernel and enhances the nonlinear representation through the GELU activation function with BatchNorm. (2) Channel interaction is when the spatial features of each channel are extracted independently using depthwise convolution; expression ability. (3) Channel interaction enhancement: To avoid the problem of weakening inter-channel relationships caused by deep convolution, pointwise convolution (1 × 1 convolution) is introduced to dynamically fuse cross-channel information and reconstruct the channel dependency. Further, the contextual correlation between occluded and visible regions is explicitly modeled by global channel feature fusion of multi-branch outputs through a two-layer fully connected network. Finally, the channel-wise exponential scaling (ES) features are multiplied element-by-element with the original input features to achieve feature compensation and semantic enhancement of occluded targets. This design significantly improves target recognizability in complex occlusion scenarios through multi-granularity local sensing and channel-wise adaptive feature selection mechanism.

Targeting the key challenges of minute target dimensions, low signal-to-noise ratio, and background clutter interference in infrared small target detection, MultiSEAM delivers performance advancements through multi-scale context fusion and channel attention co-optimization. In the spatial dimension, CSMM branches of different sizes focus on local details (e.g., patch = 6 to capture edge texture) and global context (e.g., patch = 8 to model long-range dependency) to form complementary multi-scale feature representations; in the channel dimension, the feature recalibration mechanism based on the attention weight (exponential expansion after Sigmoid activation) suppresses the background thermal noise channel and, at the same time, reinforces the key channel response related to the target. This feature enables robust detection performance in low illumination and strong occlusion infrared scenes, providing an efficient solution for real-time detection of small targets in complex environments.

3.3. *C2f_DWR*

Many recent methods have designed specialized backbones for real-time target detection tasks, of which the design of the receptive field is an important part [28,29]. Generally, these approaches seek a wide receptive field to acquire more contextual information, thereby enhancing feature representation. However, in practice, this may lead to inefficient feature extraction. Determining the appropriate size of the receptive field is crucial for improving the efficiency of feature extraction, and the requirement of the receptive field size varies at different stages of the network. With the enhancement of feature semantic representation, larger receptive fields are demanded at higher network stages, particularly for infrared small target detection. This study introduces novel improvements to the core modules of the YOLOv8 network, aiming to efficiently acquire multi-scale contextual information in infrared small target detection tasks. As the critical feature extraction component in target detection networks, the original C2f module realizes cross-layer feature fusion through a cascaded bottleneck structure, but its fixed-size 3 × 3 standard convolution has limitations in multi-scale feature extraction in complex contexts. Therefore, our method adopts DWR (dilation-wise residual) instead of bottleneck to compose C2f_DWR for IR small target detection.

As depicted in Figure 6, the DWR module employs an innovative dual-phase feature extraction strategy to resolve the critical challenge of receptive field expansion while preserving computational efficiency for real-time detection tasks. This design enables comprehensive contextual feature integration without compromising inference speed. This approach decomposes traditional single-step multi-scale feature extraction into two sequential processes: region residualization and semantic residualization. First, region residualization involves generating compact feature maps with distinct regional representations through 3 × 3 convolution integrated with batch normalization (BN) and ReLU activation, which establishes the foundation for morphological filtering in the subsequent step. Subsequent semantic residualization refers to semantically based morphological filtering using a depth-separable convolution of a single desired receptive field for each region feature map, avoiding unnecessary redundant receptive fields. This module employs a two-phase residual feature extraction approach (regional residualization-semantic residualization) to effectively enhance the efficiency of multi-scale information acquisition.

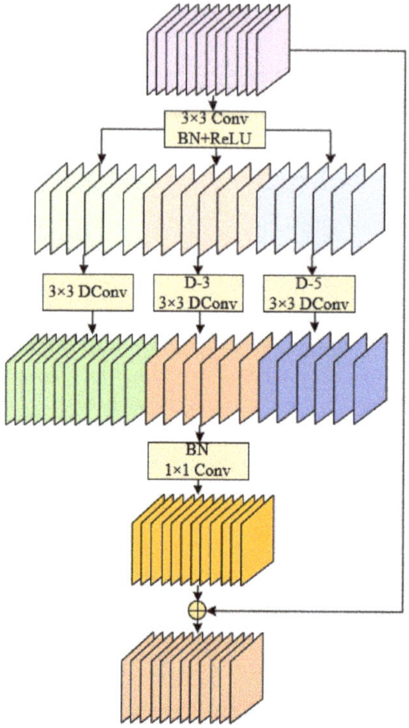

Figure 6. The network structure of the DWR module, where Conv represents convolution, DConv signifies depthwise convolution, D-n denotes dilated convolution with a dilation rate of n, and the circled '+' indicates an addition operation.

4. Experiments and Results

This section first introduces the evaluation metrics and implementation details. To demonstrate the superiority of our approach, we then conduct comparative evaluations on both the DroneVehicle and DroneRoadVehicles benchmarks. Finally, extensive ablation studies validate the individual contributions of the MultiSEAM and C2f_DWR modules.

4.1. Evaluation Metrics

Target detection algorithms employ multiple evaluation metrics that assess algorithm performance from diverse perspectives. The evaluation employs five key metrics: detection accuracy (precision, recall), overall performance (mean average precision), and computational efficiency (model parameters, FLOPs).

4.1.1. Precision and Recall

The confusion matrix, a tabular format for evaluating classification model performance [30], compares model predictions against true labels and categorizes them into four scenarios: true positives (TP), true negatives (TN), false positives (FP), and false negatives (FN). As demonstrated in Table 2, the matrix rows denote actual categories, while columns represent predicted categories.

Table 2. Confusion matrix.

Actual/Predicted	Positive	Negative
Positive	True Positive (TP)	False Negative (FN)
Negative	False Positive (FP)	True Negative (TN)

Precision quantifies the model's ability to correctly identify positive instances, measured as the ratio of true positives (TP) to all predicted positives (TP + FP). In object detection, a prediction is considered valid only when the detected bounding box meets specific overlap criteria (e.g., IoU threshold) with the ground-truth annotation. The mathematical formulation is given by Equation (1).

$$\mathrm{P} = \frac{\mathrm{TP}}{\mathrm{TP} + \mathrm{FP}} \tag{1}$$

Recall measures the model's coverage of positive instances, calculated as the ratio of true positives (TP) to all actual positives (TP + FN). For object detection, a ground-truth bounding box is considered successfully recalled if it matches with any predicted box meeting the predefined Intersection-over-Union (IoU) criterion. The formal definition is provided in Equation (2).

$$\mathrm{R} = \frac{\mathrm{TP}}{\mathrm{TP} + \mathrm{FN}} \tag{2}$$

The F1 score, defined as the harmonic mean of precision and recall, provides a balanced assessment of classification performance by equally weighting both Type I (false positives) and Type II (false negatives) errors. The calculation formula is provided in (3). Typically ranging between 0 and 1, the F1 score signifies an ideal classifier at 1 and the poorest performance at 0. This metric is especially valuable in scenarios involving imbalanced categories, as it accounts for both false positives and false negatives rather than relying solely on accuracy. In many cases, a trade-off often exists between precision and recall, and the F1 score aids in identifying an optimal balance where the classifier demonstrates robust performance across both metrics.

$$\mathrm{F1} = \frac{2 \times \mathrm{R} \times \mathrm{P}}{\mathrm{R} + \mathrm{P}} \tag{3}$$

4.1.2. mAP50 and mAP50-95

Average precision (AP) and its extension, mean average precision (MAP), constitute fundamental performance measures for object detection and semantic segmentation tasks, providing rigorous quantification of localization accuracy across all relevant object categories. Average precision (AP) serves as a fundamental performance metric in object detection, obtained by integrating precision values across all recall levels for individual object categories. This evaluation is conducted through the precision–recall (PR) curve analysis, where precision measurements are systematically recorded at incrementally varying recall thresholds. The definitive AP score is subsequently derived by computing the total area beneath this PR curve (commonly termed AUC-PR to distinguish it from ROC curve analysis), with this integration process effectively consolidating both detection accuracy and completeness into a single quantitative measure. The specific formula is

$$\mathrm{AP} = \int_0^1 p(r)\,\mathrm{d}r \tag{4}$$

where $p(r)$ denotes the precision at recall level r.

The mean average precision (mAP) serves as the primary evaluation metric by computing the arithmetic mean of category-wise AP values, thereby offering a holistic measure of model performance across all target classes. This aggregated metric is mathematically expressed in Equation (5).

$$\mathrm{mAP} = \frac{1}{n}\sum_{i=1}^{n}\int_{0}^{1} p(r)\,\mathrm{d}r \tag{5}$$

4.1.3. Parameters and GFLOPs

Beyond task-specific performance metrics (e.g., accuracy), model efficiency metrics like parameter count and operational complexity must also be considered. The parameter count, denoting the number of model parameters, reflects the memory required for storage; operational complexity is typically measured in FLOPs (floating point operations), characterizing the computational resources needed for model inference. It is noted that FLOPs refer to the count of floating-point operations, commonly quantified in GFLOPs (10^9).

4.2. Experimental Setup

We use YOLOv8 as our baseline, and the method is implemented in PyTorch 1.12 using a GPU that is an NVIDIA GeForce RTX 3080Ti. In the experiments, the batch size is set to 8, and the training optimization algorithm is Stochastic Gradient Descent (SGD) with an initial learning rate of 0.01 and momentum and weight decay of 0.9 and 0.0005, respectively.

4.3. Comparison Experiments

To validate the efficacy of YOLO-HVS in infrared small target detection under complex conditions, we conduct comparative experiments on two datasets: the public DroneVehicle dataset and our self-built DroneRoadVehicles dataset (containing 1028 long-range infrared images with occlusions). Four methods are evaluated: Faster R-CNN [31], YOLOv8, YOLOv11 [32], and our YOLO-HVS with MultiSEAM and C2f_DWR modules.

As shown in Table 3 (DroneVehicle dataset), YOLO-HVS achieves state-of-the-art performance with 83.4% mAP50 (1.1% higher than YOLOv8) and 62.5% mAP50-95, while maintaining real-time efficiency (41 FPS). Notably, the proposed MultiSEAM effectively suppresses background interference, reflected in the highest recall (0.787) and F1-score (0.797). Although YOLOv11 achieves lower computational costs (6.3 GFLOPs), its accuracy (82.8% mAP50) is inferior to our method, demonstrating the superiority of our multi-scale feature fusion strategy.

On the DroneRoadVehicles dataset (Table 4), YOLO-HVS further validates its robustness in occlusion scenarios, reaching 97.8% mAP50 (0.7% improvement over YOLOv8) and 64% mAP50-95. The C2f_DWR module's dilation-wise residual structure enhances multi-scale target detection, as evidenced by the highest recall (0.962) despite complex backgrounds. Visualization results confirm that YOLO-HVS outperforms competitors in low signal-to-noise ratio scenarios.

Table 3. Comparative Experiments on DroneVehicle Dataset.

Method	P	R	F1	mAP50	mAP50-95	GFLOPs	Params (M)	FPS
Faster R-CNN	0.73	0.59	0.653	0.674	–	207	41	–
YOLOv8	0.804	0.778	0.791	0.823	0.616	8.1	3.1	26
YOLOv11	**0.809**	0.782	0.79	0.828	0.622	6.3	2.6	50
Ours	0.807	**0.787**	**0.797**	**0.834**	**0.625**	8.2	5.4	41

Table 4. Performance Comparison on DroneRoadVehicles Dataset.

Method	P	R	F1	mAP50	mAP50-95
Faster R-CNN	0.887	0.78	0.83	0.823	-
YOLOv8	0.931	0.96	**0.945**	0.971	0.63
YOLOv11	**0.933**	0.957	**0.945**	0.975	0.638
Ours	0.926	**0.962**	0.944	**0.978**	**0.64**

4.4. Ablation Experiments

The proposed framework integrates two key components for infrared small target detection: (1) the MultiSEAM attention mechanism for cross-scale feature enhancement and (2) the C2f_DWR module for dynamic receptive field adaptation. Systematic ablation studies (Table 5) quantitatively demonstrate their individual contributions to detection performance improvement.

As demonstrated in Table 5, on the DroneVehicle dataset, MultiSEAM and C2f_DWR improve mAP50 by 0.4% (from 0.823 to 0.827) and 0.2% (from 0.823 to 0.825), respectively, compared to baseline YOLOv8. The mAP50-95 metric shows corresponding improvements of 0.3% (0.616→0.619) and 0.2% (0.616→0.618). Our proposed method achieves more significant enhancements on the custom dataset (ours) with combined modules delivering 0.7% mAP50 improvement (0.971→0.978).

Notably, the C2f_DWR module demonstrates dual benefits—while improving detection accuracy, it reduces computational complexity by 3.7% in GFLOPs (8.1→7.8) and model parameters by 9.7% (3.1 M→2.8 M), confirming its lightweight characteristics. This enables our model to maintain baseline efficiency while achieving superior detection performance.

Table 5. MultiSEAM and C2f_DWR ablation experiments.

Dataset	MultiSEAM	C2f_DWR	mAP50	mAP50-95	GFLOPs	Params (M)
DroneVehicle			0.823	0.616	8.1	3.1
	✓		0.827	0.619	8.6	5.6
		✓	0.825	0.618	7.8	2.8
	✓	✓	**0.834**	**0.625**	8.2	5.4
DroneRoadVehicles			0.971	0.630	8.1	3.1
	✓		0.974	0.639	8.6	5.6
		✓	0.972	**0.642**	7.8	2.8
	✓	✓	**0.978**	0.640	8.2	5.4

4.5. Visualization of Results Analysis

To qualitatively evaluate the performance superiority of our approach in detecting infrared small targets under challenging conditions—particularly for multi-scale objects amidst complex background clutter—we conducted visual comparisons using representative samples from the DroneRoadVehicles test set. As demonstrated in Figures 7–9, each case study presents: (a) the original infrared input, (b) detection outputs from the YOLOv8 baseline, and (c) predictions generated by our YOLO-HVS framework.

First, in the scenario shown in Figure 7, the core challenge lies in effectively detecting a small-scale vehicle target that is heavily interfered with by road thermal radiation noise and tree trunks (as indicated by the green bounding box in the original Figure 7a). The baseline model YOLOv8 (Figure 7b) failed to identify this target, exhibiting significant false negatives. This reflects the limitations of the original model in handling small targets and suppressing complex background noise. In contrast, our method (Figure 7c) successfully detected the target. This significant improvement is clearly attributed to the integrated MultiSEAM attention mechanism. This module can adaptively focus on critical small target regions, enhancing the model's ability to extract and distinguish weak target features in information-sparse and noise-intensive scenes, effectively addressing the issue of small target false negatives.

Furthermore, we examined the model's performance in detecting standard-sized targets and its confidence output. Figure 8 shows a scene containing clearly visible vehicle targets. Both YOLOv8 (Figure 8b) and our method (Figure 8c) successfully detected these targets, indicating that both methods possess foundational capabilities in identifying and locating standard-sized targets. However, a detailed comparison reveals a key difference: our method's detection results are generally accompanied by significantly higher confidence levels. As shown in Figure 8c, the confidence values of the vehicle detection boxes are notably and consistently higher than those of YOLOv8 in Figure 8b. This strongly suggests that the C2f_DWR module not only optimizes feature extraction efficiency (as demonstrated by the lightweight characteristics proven in the ablation experiments mentioned earlier), but more importantly, it generates more robust and discriminative feature representations. This enhanced feature representation directly translates into higher prediction confidence for the detector, improving the certainty of the model's output results, which is crucial for subsequent application scenarios (such as early warning and tracking).

Finally, to validate the model's generalization ability and consistent improvement for typical target categories, Figure 9 focuses on analyzing a truck detection scenario. As a target with distinct size and appearance features within the vehicle category, the detection results for the truck further validate the model's applicability. Both YOLOv8 (Figure 9b) and the proposed method (Figure 9c) successfully detected the truck target. Building on the observations from the previous scenario, the confidence scores of the detection bounding boxes output by this method are again significantly higher than those of the baseline model. This case not only reinforces the conclusion of improved confidence scores on another target category but also demonstrates that the combination of the MultiSEAM and C2f_DWR modules has universally enhanced the model's overall feature extraction capabilities and confidence discrimination, rather than being effective only on specific target types.

Quantitative analysis (see Table 4) further supports the aforementioned visualization conclusions: on a specific test subset that specifically reflects complex background interference, YOLO-HVS improves the key metric mAP50-95 from 0.63 to 0.64 compared to the baseline YOLOv8. Combining visualization and quantitative results, YOLO-HVS enhances small object detection capabilities by integrating the MultiSEAM module and by utilizing the C2f_DWR module to optimize feature extraction and confidence generation, significantly improving the accuracy and reliability of infrared small target detection under complex background noise and multi-scale targets. This is of great value for enhancing the perception robustness and decision confidence of unmanned aerial vehicle (UAV) nighttime road monitoring systems in practical applications.

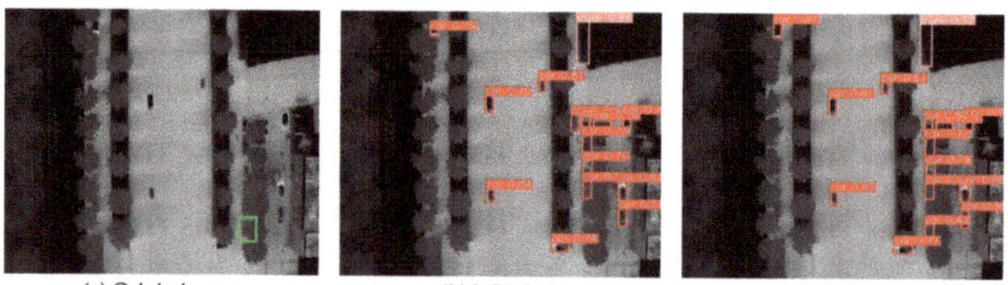

(a) Origin image (b) YOLOv8 (c) Ours

Figure 7. Comparison of small-scale car target detection.

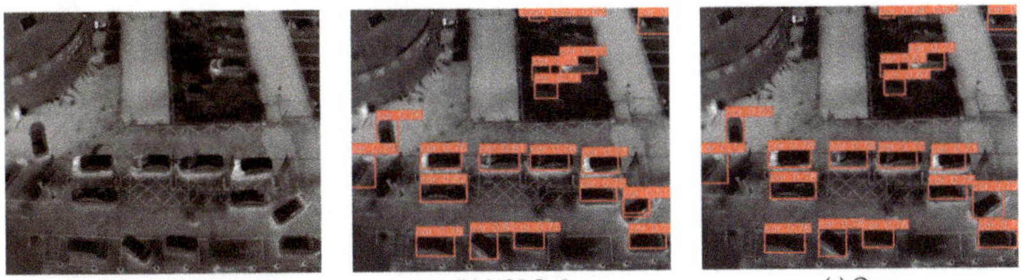

(a) Original image (b) YOLOv8 (c) Ours

Figure 8. Comparison of standard-scale car target detection.

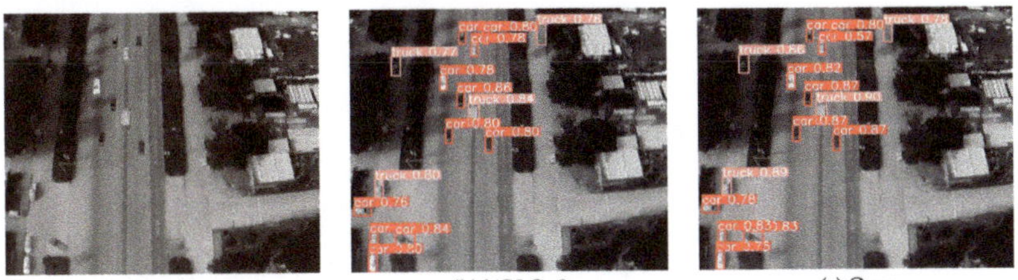

(a) Original image (b) YOLOv8 (c) Ours

Figure 9. Truck target detection comparison.

5. Conclusions

This paper presents YOLO-HVS, an enhanced YOLOv8-based algorithm specifically designed to address the problems of background interference and multi-scale feature extraction in infrared small target detection. By integrating the attention mechanism of the human visual system and the multi-scale perceptual characteristics, we innovatively designed the MultiSEAM attention module and the C2f_DWR feature extraction module. MultiSEAM significantly improves the target recognition ability in complex occlusion scenes by joint modeling of multi-scale spatial channels, while the C2f_DWR module effectively enlarges the range of the sensory field by the region-semantic dual residual structure. MultiSEAM significantly improves the target recognition ability in complex occlusion scenes through joint modeling of granularity spatial channels, while the C2f_DWR module effectively enlarges the perceptual field range through the region-semantic dual residual structure, realizing the efficient extraction of multi-scale features. The constructed DroneRoadVehicles dataset covers UAV multi-height shooting scenarios, which provides important data support for infrared small target detection research. Comparison experiments on DroneVehicle and self-constructed datasets show that the proposed method improves the mAP50-95 metrics by 0.9% and 1.0%, respectively, while keeping the model lightweight (parametric count of 7.6M, GFLOPs of 10.8), and ablation experiments further validate the effectiveness of the modules. Future work will explore the dynamic sensory field adjustment mechanism to further enhance the adaptability of the algorithm in extreme scale change scenarios.

Author Contributions: Conceptualization, X.W. and Y.S.; methodology, X.W. and Q.H.; software, H.H.; validation, X.W., Y.S. and S.N.; formal analysis, X.W.; investigation, X.W.; resources, Q.H.; data curation, X.W.; writing—original draft preparation, X.W.; writing—review and editing, X.W. All authors have read and agreed to the published version of the manuscript.

Funding: This research received no external funding.

Institutional Review Board Statement: Not applicable.

Informed Consent Statement: This study utilized infrared imaging data for vehicle detection, which did not involve human participants. Therefore, ethical approval and informed consent are not applicable.

Data Availability Statement: The original contributions presented in this study are included in the article. Further inquiries can be directed to the corresponding author.

Conflicts of Interest: The authors declare no conflict of interest.

References

1. Zhao, M.; Li, W.; Li, L.; Hu, J.; Ma, P.; Tao, R. Single-Frame Infrared Small-Target Detection: A survey. *IEEE Geosci. Remote Sens. Mag.* **2022**, *10*, 87–119. [CrossRef]
2. Zhang, Q.; Zhou, L.; An, J. Real-Time Recognition Algorithm of Small Target for UAV Infrared Detection. *Sensors* **2024**, *24*, 3075. [CrossRef] [PubMed]
3. Gao, C.; Meng, D.; Yang, Y.; Wang, Y.; Zhou, X.; Hauptmann, A.G. Infrared Patch-Image Model for Small Target Detection in a Single Image. *IEEE Trans. Image Process.* **2013**, *22*, 4996–5009. [CrossRef]
4. Rawat, S.S.; Verma, S.K.; Kumar, Y. Review on recent development in infrared small target detection algorithms. *Procedia Comput. Sci.* **2020**, *167*, 2496–2505. [CrossRef]
5. Ju, M.; Luo, J.; Liu, G.; Luo, H. ISTDet: An efficient end-to-end neural network for infrared small target detection. *Infrared Phys. Technol.* **2021**, *114*, 103659. [CrossRef]
6. Wang, K.; Li, S.; Niu, S.; Zhang, K. Detection of Infrared Small Targets Using Feature Fusion Convolutional Network. *IEEE Access* **2019**, *7*, 146081–146092. [CrossRef]
7. Jeon, Y.; Chang, W.; Jeong, S.; Han, S.; Park, J. A Bayesian convolutional neural network-based generalized linear model. *Biometrics* **2024**, *80*, ujae057. [CrossRef] [PubMed]
8. Zhang, J.; Liu, C.; Wang, B.; Chen, C.; He, J.; Zhou, Y.; Li, J. An infrared pedestrian detection method based on segmentation and domain adaptation learning. *Comput. Electr. Eng.* **2022**, *99*, 107781. [CrossRef]
9. Redmon, J.; Divvala, S.; Girshick, R.; Farhadi, A. You Only Look Once: Unified, Real-Time Object Detection. In Proceedings of the 2016 IEEE Conference on Computer Vision and Pattern Recognition (CVPR), Los Alamitos, CA, USA, 27–30 June 2016; pp. 779–788.
10. Redmon, J.; Farhadi, A. YOLO9000: Better, Faster, Stronger. In Proceedings of the 2017 IEEE Conference on Computer Vision and Pattern Recognition (CVPR), Honolulu, HI, USA, 21–26 July 2017; pp. 6517–6525. [CrossRef]
11. Yue, T.; Lu, X.; Cai, J.; Chen, Y.; Chu, S. YOLO-MST: Multiscale deep learning method for infrared small target detection based on super-resolution and YOLO. *Opt. Laser Technol.* **2025**, *187*, 112835. [CrossRef]
12. Akshatha, K.R.; Karunakar, A.K.; Shenoy, S.B.; Pai, A.K.; Nagaraj, N.H.; Rohatgi, S.S. Human Detection in Aerial Thermal Images Using Faster R-CNN and SSD Algorithms. *Electronics* **2022**, *11*, 1151. [CrossRef]
13. Han, J.; Liang, K.; Zhou, B.; Zhu, X.; Zhao, J.; Zhao, L. Infrared Small Target Detection Utilizing the Multiscale Relative Local Contrast Measure. *IEEE Geosci. Remote Sens. Lett.* **2018**, *15*, 612–616. [CrossRef]
14. Han, J.; Ma, Y.; Huang, J.; Mei, X.; Ma, J. An Infrared Small Target Detecting Algorithm Based on Human Visual System. *IEEE Geosci. Remote Sens. Lett.* **2016**, *13*, 452–456. [CrossRef]
15. Han, J.; Ma, Y.; Zhou, B.; Fan, F.; Liang, K.; Fang, Y. A Robust Infrared Small Target Detection Algorithm Based on Human Visual System. *IEEE Geosci. Remote Sens. Lett.* **2014**, *11*, 2168–2172. [CrossRef]
16. Guo, M.H.; Xu, T.X.; Liu, J.J.; Liu, Z.N.; Jiang, P.T.; Mu, T.J.; Zhang, S.H.; Martin, R.R.; Cheng, M.M.; Hu, S.M. Attention mechanisms in computer vision: A survey. *Comput. Vis. Media* **2022**, *8*, 331–368. [CrossRef]
17. Du, S.; Zhang, B.; Zhang, P.; Xiang, P.; Xue, H. FA-YOLO: An Improved YOLO Model for Infrared Occlusion Object Detection under Confusing Background. *Wirel. Commun. Mob. Comput.* **2021**, *2021*. [CrossRef]
18. Deng, H.; Zhang, Y. FMR-YOLO: Infrared Ship Rotating Target Detection Based on Synthetic Fog and Multiscale Weighted Feature Fusion. *IEEE Trans. Instrum. Meas.* **2024**, *73*,5001717. [CrossRef]
19. Zhang, T.; Li, L.; Cao, S.; Pu, T.; Peng, Z. Attention-Guided Pyramid Context Networks for Detecting Infrared Small Target Under Complex Background. *IEEE Trans. Aerosp. Electron. Syst.* **2023**, *59*, 4250–4261. [CrossRef]
20. Song, Z.; Yan, Y.; Cao, Y.; Jin, S.; Qi, F.; Li, Z.; Lei, T.; Chen, L.; Jing, Y.; Xia, J.; et al. An infrared dataset for partially occluded person detection in complex environment for search and rescue. *Sci. Data* **2025**, *12*, 300. [CrossRef] [PubMed]
21. Wang, Y.; Tian, Y.; Liu, J.; Xu, Y. Multi-Stage Multi-Scale Local Feature Fusion for Infrared Small Target Detection. *Remote Sens.* **2023**, *15*, 4506. [CrossRef]
22. Wang, Y.; Jiang, P.; Pan, N. Infrared small target detection based on local significance and multiscale. *Digit. Signal Process.* **2024**, *155*, 104721. [CrossRef]
23. Sun, Y.; Cao, B.; Zhu, P.; Hu, Q. Drone-Based RGB-Infrared Cross-Modality Vehicle Detection Via Uncertainty-Aware Learning. *IEEE Trans. Circuits Syst. Video Technol.* **2022**, *32*, 6700–6713. [CrossRef]
24. Wang, J.; Teng, X.; Li, Z.; Yu, Q.; Bian, Y.; Wei, J. VSAI: A Multi-View Dataset for Vehicle Detection in Complex Scenarios Using Aerial Images. *Drones* **2022**, *6*, 161. [CrossRef]

25. Varghese, R.; Sambath, M. YOLOv8: A Novel Object Detection Algorithm with Enhanced Performance and Robustness. In Proceedings of the 2024 International Conference on Advances in Data Engineering and Intelligent Computing Systems (ADICS), Chennai, India, 18–19 April 2024; pp. 1–6.
26. Yu, Z.; Huang, H.; Chen, W.; Su, Y.; Liu, Y.; Wang, X. YOLO-FaceV2: A scale and occlusion aware face detector. *Pattern Recognit.* **2024**, *155*, 110714. [CrossRef]
27. Muhammad, W.; Aramvith, S.; Onoye, T. Multi-scale Xception based depthwise separable convolution for single image super-resolution. *PLoS ONE* **2021**, *16*, e0249278. [CrossRef]
28. Araujo, A.; Norris, W.; Sim, J. Computing receptive fields of convolutional neural networks. *Distill* **2019**, *4*. [CrossRef]
29. Chen, Q.; Li, C.; Ning, J.; Lin, S.; He, K. GMConv: Modulating Effective Receptive Fields for Convolutional Kernels. *IEEE Trans. Neural Netw. Learn. Syst.* **2025**, *36*, 6669–6678. [CrossRef]
30. Markoulidakis, I.; Markoulidakis, G. Probabilistic Confusion Matrix: A Novel Method for Machine Learning Algorithm Generalized Performance Analysis. *Technologies* **2024**, *12*, 113. [CrossRef]
31. Ren, S.; He, K.; Girshick, R.; Sun, J. Faster R-CNN: Towards Real-Time Object Detection with Region Proposal Networks. *IEEE Trans. Pattern Anal. Mach. Intell.* **2017**, *39*, 1137–1149. [CrossRef]
32. Khanam, R.; Hussain, M. YOLOv11: An Overview of the Key Architectural Enhancements. *arXiv* **2024**, arXiv:2410.17725.

biomimetics

MDPI

Communication

The Effects of Turbulent Biological Tissue on Adjustable Anomalous Vortex Laser Beam

Yiqun Zhang [1,2], Wu Wang [3], Xiaokun Ding [4,5], Liyu Sun [2], Zhenyang Qian [2], Huilin Jiang [1], Yansong Song [1,2] and Runwei Ding [2,*]

1 School of Optoelectronic Engineering, Changchun University of Science and Technology, Changchun 130013, China; zhangyq@pcl.ac.cn (Y.Z.)
2 Peng Cheng Laboratory, Shenzhen 518055, China
3 North Navigation Control Technology Co., Ltd., Beijing 100072, China
4 AVIC Flight Automatic Control Research Institute, Northwestern Polytechnical University, Xi'an 710072, China
5 AVIC Flight Automatic Control Research Institute, Xi'an 710065, China
* Correspondence: dingrw@pcl.ac.cn

Abstract

In this work, we present a new partially coherent adjustable anomalous vortex laser beam (PCAAVLB) and introduce it into turbulent biological tissue. The equation of such PCAAVLB in turbulent biological tissue is obtained. By numerical analysis, the evolution of the intensity of such PCAAVLB in turbulent biological tissue is analyzed. It is found that the PCAAVLB in biological tissue can lose its ring shape and become a Gaussian beam, and a PCAAVLB with smaller topological charge M or coherence length σ will evolve into a Gaussian profile faster. The PCAAVLB in turbulent biological tissue with a smaller small-length-scale factor l_0 or larger fractal dimension D will evolve into a Gaussian profile faster and have a larger intensity as z increases. The results may have potential applications in sensing under biological tissue environments and laser imaging in biology.

Keywords: laser; interdisciplinary research; intensity; biological tissue

1. Introduction

The properties of tissue affect the application of lasers in medicine, and the optical properties of tissue have been investigated [1]. To explore the applications of lasers in biological tissues, the influences of tissue turbulence on optical waves have attracted attention [2], as have the properties (intensity and beam spread et al.) of light in biological tissues. In addition, the intensity of various lasers in turbulent biological tissue has been studied, such as the hollow Gaussian beam [3], partially coherent beam [4], vortex beam [5], Hermite–Gaussian correlated beam [6], rectangular multi-Gaussian beam [7], and Gaussian–Schell model vortex beam [8]. The coherence properties of laser in tissue are still analyzed [9]. Moreover, the coupling efficiency to fiber [10] and adaptive optics correction in tissues [11] are also investigated. From the above references, one sees that different types of laser beams may have potential applications in tissues.

A vortex is a signature of natural systems and is observed in fluids, smoke rings, and tornados [12]. The high speed of the peregrine falcon can be enhanced by the vortices emanating from the frontal and dorsal regions [13]. The trailing-edge vortices have been found in the flight of mosquitoes [14]. Moreover, the vortex has also been introduced into the light, and the double vortex beams have also been investigated [15,16]. The

 https://doi.org/10.3390/biomimetics10070461

intensity pattern of a laser usually has a Gaussian profile. However, the intensity pattern of light can be modulated, and an anomalous hollow beam can show a unique intensity profile [17]. And the propagation of such AHB has been studied [18–20]. Such an AHB has also been extended into a partially coherent beam [21–23]. Recently, a new adjustable anomalous vortex beam (AAVB) has been introduced [24]; the properties of such AAVB can be modulated by adjustable parameters. It will be interesting if this adjustable beam is introduced into biological tissue.

The vortex in nature [12], birds [13] and mosquitoes [14], shows unique properties. In this work, the AAVB is extended into a partially coherent beam, and a new partially coherent adjustable anomalous vortex laser beam (PCAAVLB), which can be modulated by the adjustable parameters, is introduced. The intensity of the PCAAVLB can be controlled by adjustable parameters, while the intensity of the multi-Gaussian correlated AHB has the ring symmetry in Reference [23]. This PCAAVLB is introduced into the tissues, and the intensity equation of such PCAAVLB in turbulent biological tissue is derived. Based on the derived equations, the effects of turbulent biological tissue on the intensity of such PCAAVLB are analyzed.

2. Propagation of PCAAVLB in Turbulent Biological Tissue

When the PCAAVLB propagates along the z-axis, the intensity of PCAAVLB in turbulent biological tissue is described by the Huygens–Fresnel integral as follows [3–8]:

$$\begin{aligned} I(\boldsymbol{\rho}, z) &= \frac{k^2}{4\pi^2 z^2} \int d\mathbf{r}_1 d\mathbf{r}_2 W_0(\mathbf{r}_1, \mathbf{r}_2) \\ &\times \exp\left[-\frac{ik}{2z}(\boldsymbol{\rho} - \mathbf{r}_1)^2 + \frac{ik}{2z}(\boldsymbol{\rho} - \mathbf{r}_2)^2\right] \langle \exp[\psi(\mathbf{r}_1, \boldsymbol{\rho}) + \psi^*(\mathbf{r}_2, \boldsymbol{\rho})] \rangle \end{aligned} \tag{1}$$

where $I(\boldsymbol{\rho}, z)$ represents the intensity; $W_0(\mathbf{r}_1, \mathbf{r}_2)$ is the cross spectral density (CSD) of laser at $z = 0$; $k = 2\pi/\lambda$ is the wavenumber with λ is the wavelength in vacuum; $\mathbf{r} = (x, y)$ and $\boldsymbol{\rho} = (\rho_x, \rho_y)$ are position vectors at $z = 0$ and z, respectively; i is imaginary unit; $\psi(\mathbf{r}, \boldsymbol{\rho})$ is the phase perturbation of biological tissue.

$$\langle \exp[\psi(\mathbf{r}_1, \boldsymbol{\rho}) + \psi^*(\mathbf{r}_2, \boldsymbol{\rho})] \rangle = \exp\left[-\frac{(x_1 - x_2)^2 + (y_1 - y_2)^2}{\Lambda^2}\right] \tag{2}$$

The Λ is the coherence length of a spherical wave in turbulent biological tissue and which can be written as follows [10]:

$$\Lambda = \left[\frac{\pi^{1/2} k^2 S z}{3 \times 2^{(7-D)/2} l_c} \Gamma\left(\frac{D}{2}\right) U\left(2; \frac{7}{6}; \frac{l_0^2}{8 \ln(2) l_c^2}\right)\right]^{-1/2} \tag{3}$$

where S denotes the strength coefficient of biological tissues; D denotes the fractal dimension; l_c denotes the characteristic length of heterogeneity; l_0 denotes small length-scale factor; $\Gamma(\cdot)$ denotes the gamma function; $U(\cdot)$ denotes the confluent hypergeometric function of the second kind.

The electric field of AAVB at $z = 0$ is written as follows [24,25]:

$$E(\mathbf{r}, 0) = \left(-a + c_x \frac{x^2}{w_x^2} + c_y \frac{y^2}{w_y^2}\right) \exp\left(-\frac{x^2}{w_x^2} - \frac{y^2}{w_y^2}\right) (x + iy)^M \tag{4}$$

where w_x and w_y are the beam width of the Gaussian part; M is a topological charge; and a, c_x, and c_y are adjustable parameters.

Considering the coherence theory [26], the CSD of a PCAAVLB is written as follows:

$$W_0(\mathbf{r}_1,\mathbf{r}_2) = \left(-a + c_x\frac{{x_1}^2}{w_x^2} + c_y\frac{{y_1}^2}{w_y^2}\right)\exp\left(-\frac{{x_1}^2}{w_x^2} - \frac{{y_1}^2}{w_y^2}\right)(x_1 + iy_1)^M \\ \times\left(-a + c_x\frac{{x_2}^2}{w_x^2} + c_y\frac{{y_2}^2}{w_y^2}\right)\exp\left(-\frac{{x_2}^2}{w_x^2}-,\frac{{y_2}^2}{w_y^2}\right)(x_2 + iy_2)^M \\ \times\exp\left[-\frac{(x_1-x_2)^2}{2\sigma_x^2} - \frac{(y_1-y_2)^2}{2\sigma_y^2}\right] \tag{5}$$

where σ_x and σ_y are the coherence length along x-axis and y-axis. The intensity of PCAAVLB can be modulated by the adjustable parameters in Equation (5), and this beam is newly introduced, which is different from the multi-Gaussian correlated partially coherent anomalous hollow beam in Reference [23].

We next consider the following equations [27]:

$$(x+iy)^M = \sum_{l=0}^{M}\frac{M!i^l}{l!(M-l)!}x^{M-1}y^l \tag{6}$$

$$\int_{-\infty}^{+\infty} x^n\exp\left(-px^2+2qx\right)dx = n!\exp\left(\frac{q^2}{p}\right)\left(\frac{q}{p}\right)^n\sqrt{\frac{\pi}{p}}\sum_{l=0}^{[\frac{n}{2}]}\frac{1}{l!(n-2l)!}\left(\frac{p}{4q^2}\right)^l \tag{7}$$

By substituting Equation (5) into Equation (1), the intensity of such PCAAVLB in turbulent biological tissue is derived as follows:

$$I(\rho,z) = \frac{k^2}{4\pi^2z^2}\sum_{m=0}^{M}\frac{M!i^m}{m!(M-m)!}\sum_{l=0}^{M}\frac{M!i^n}{n!(M-n)!} \\ \left(a^2I_1 - \frac{ac_x}{w_x^2}I_2 - \frac{ac_y}{w_y^2}I_3 - \frac{ac_x}{w_x^2}I_4 - \frac{ac_y}{w_y^2}I_5 + \frac{c_x^2}{w_x^4}I_6 + \frac{c_xc_y}{w_x^2w_y^2}I_7 + \frac{c_xc_y}{w_x^2w_y^2}I_8 + \frac{c_y^2}{w_y^4}I_9\right) \tag{8}$$

with

$$I_1 = \sqrt{\frac{\pi}{a_x}}(M-m)!\left(\frac{1}{a_x}\right)^{M-m}\exp\left[\frac{1}{a_x}\left(\frac{ik}{2z}\rho_x\right)^2\right] \\ \sum_{k=0}^{[\frac{M-m}{2}]}\frac{1}{k!(M-m-2k)!}\left(\frac{a_x}{4}\right)^{k[M-m-2k]}\sum_{l=0}\frac{(M-m-2k)!}{l!(M-m-2k-l)!}\left(\frac{ik}{2z}\rho_x\right)^{M-m-2k-l} \\ \left(\frac{1}{2\sigma_x^2}+\frac{1}{\Lambda^2}\right)^l\sqrt{\frac{\pi}{b_x}}2^{-(M-n+l)}i^{M-n+l}\exp\left(\frac{c_x^2}{b_x}\right)\left(\frac{1}{b_x}\right)^{0.5(M-n+l)}H_{M-n+l}\left(\frac{ic_x}{\sqrt{b_x}}\right) \\ \sqrt{\frac{\pi}{a_y}}m!\left(\frac{1}{a_y}\right)^m\exp\left[\frac{1}{a_y}\left(\frac{ik}{2z}\rho_y\right)^2\right] \\ \sum_{t=0}^{[\frac{m}{2}]}\frac{1}{t!(m-2t)!}\left(\frac{a_y}{4}\right)^{t[m-2t]}\sum_{l\prime=0}\frac{(m-2t)!}{l\prime!(m-2t-l\prime)!}\left(\frac{ik}{2z}\rho_y\right)^{m-2t-l\prime}\left(\frac{1}{2\sigma_y^2}+\frac{1}{\Lambda^2}\right)^{l\prime} \\ \sqrt{\frac{\pi}{b_y}}2^{-(n+l\prime)}i^{n+l\prime}\exp\left(\frac{c_y^2}{b_y}\right)\left(\frac{1}{b_y}\right)^{0.5(n+l\prime)}H_{n+l\prime}\left(\frac{ic_y}{\sqrt{b_y}}\right) \tag{9}$$

$$
\begin{aligned}
I_2 = {} & \sqrt{\frac{\pi}{a_x}}(M-m+2)!\left(\frac{1}{a_x}\right)^{M-m+2} \exp\left[\frac{1}{a_x}\left(\frac{ik}{2z}\rho_x\right)^2\right] \\
& \sum_{k=0}^{[\frac{M-m+2}{2}]} \frac{1}{k!(M-m+2-2k)!}\left(\frac{a_x}{4}\right)^{k} \sum_{l=0}^{[M-m+2-2k]} \frac{(M-m+2-2k)!}{l!(M-m+2-2k-l)!}\left(\frac{ik}{2z}\rho_x\right)^{M-m+2-2k-l} \\
& \left(\frac{1}{2\sigma_x^2}+\frac{1}{\Lambda^2}\right)^{l} \sqrt{\frac{\pi}{b_x}} 2^{-(M-n+l)} i^{M-n+l} \exp\left(\frac{c_x^2}{b_x}\right)\left(\frac{1}{b_x}\right)^{0.5(M-n+l)} H_{M-n+l}\left(\frac{ic_x}{\sqrt{b_x}}\right) \\
& \sqrt{\frac{\pi}{a_y}} m! \left(\frac{1}{a_y}\right)^{m} \exp\left[\frac{1}{a_y}\left(\frac{ik}{2z}\rho_y\right)^2\right] \\
& \sum_{t=0}^{[\frac{m}{2}]} \frac{1}{t!(m-2t)!}\left(\frac{a_y}{4}\right)^{t} \sum_{l\prime=0}^{[m-2t]} \frac{(m-2t)!}{l\prime!(m-2t-l\prime)!}\left(\frac{ik}{2z}\rho_y\right)^{m-2t-l\prime} \left(\frac{1}{2\sigma_y^2}+\frac{1}{\Lambda^2}\right)^{l\prime} \\
& \sqrt{\frac{\pi}{b_y}} 2^{-(n+l\prime)} i^{n+l\prime} \exp\left(\frac{c_y^2}{b_y}\right)\left(\frac{1}{b_y}\right)^{0.5(n+l\prime)} H_{n+l\prime}\left(\frac{ic_y}{\sqrt{b_y}}\right)
\end{aligned}
\tag{10}
$$

$$
\begin{aligned}
I_3 = {} & \sqrt{\frac{\pi}{a_x}}(M-m)!\left(\frac{1}{a_x}\right)^{M-m} \exp\left[\frac{1}{a_x}\left(\frac{ik}{2z}\rho_x\right)^2\right] \\
& \sum_{k=0}^{[\frac{M-m}{2}]} \frac{1}{k!(M-m-2k)!}\left(\frac{a_x}{4}\right)^{k} \sum_{l=0}^{[M-m-2k]} \frac{(M-m-2k)!}{l!(M-m-2k-l)!}\left(\frac{ik}{2z}\rho_x\right)^{M-m-2k-l} \\
& \left(\frac{1}{2\sigma_x^2}+\frac{1}{\Lambda^2}\right)^{l} \sqrt{\frac{\pi}{b_x}} 2^{-(M-n+l)} i^{M-n+l} \exp\left(\frac{c_x^2}{b_x}\right)\left(\frac{1}{b_x}\right)^{0.5(M-n+l)} H_{M-n+l}\left(\frac{ic_x}{\sqrt{b_x}}\right) \\
& \sqrt{\frac{\pi}{a_y}} (m+2)! \left(\frac{1}{a_y}\right)^{m+2} \exp\left[\frac{1}{a_y}\left(\frac{ik}{2z}\rho_y\right)^2\right] \\
& \sum_{t=0}^{[\frac{m+2}{2}]} \frac{1}{t!(m+2-2t)!}\left(\frac{a_y}{4}\right)^{t} \sum_{l\prime=0}^{[m+2-2t]} \frac{(m+2-2t)!}{l\prime!(m+2-2t-l\prime)!}\left(\frac{ik}{2z}\rho_y\right)^{m+2-2t-l\prime} \left(\frac{1}{2\sigma_y^2}+\frac{1}{\Lambda^2}\right)^{l\prime} \\
& \sqrt{\frac{\pi}{b_y}} 2^{-(n+l\prime)} i^{n+l\prime} \exp\left(\frac{c_y^2}{b_y}\right)\left(\frac{1}{b_y}\right)^{0.5(n+l\prime)} H_{n+l\prime}\left(\frac{ic_y}{\sqrt{b_y}}\right)
\end{aligned}
\tag{11}
$$

$$
\begin{aligned}
I_4 = {} & \sqrt{\frac{\pi}{a_x}}(M-m)!\left(\frac{1}{a_x}\right)^{M-m} \exp\left[\frac{1}{a_x}\left(\frac{ik}{2z}\rho_x\right)^2\right] \\
& \sum_{k=0}^{[\frac{M-m}{2}]} \frac{1}{k!(M-m-2k)!}\left(\frac{a_x}{4}\right)^{k} \sum_{l=0}^{[M-m-2k]} \frac{(M-m-2k)!}{l!(M-m-2k-l)!}\left(\frac{ik}{2z}\rho_x\right)^{M-m-2k-l} \\
& \left(\frac{1}{2\sigma_x^2}+\frac{1}{\Lambda^2}\right)^{l} \sqrt{\frac{\pi}{b_x}} 2^{-(M-n+2+l)} i^{M-n+2+l} \exp\left(\frac{c_x^2}{b_x}\right)\left(\frac{1}{b_x}\right)^{0.5(M-n+2+l)} H_{M-n+2+l}\left(\frac{ic_x}{\sqrt{b_x}}\right) \\
& \sqrt{\frac{\pi}{a_y}} m! \left(\frac{1}{a_y}\right)^{m} \exp\left[\frac{1}{a_y}\left(\frac{ik}{2z}\rho_y\right)^2\right] \\
& \sum_{t=0}^{[\frac{m}{2}]} \frac{1}{t!(m-2t)!}\left(\frac{a_y}{4}\right)^{t} \sum_{l\prime=0}^{[m-2t]} \frac{(m-2t)!}{l\prime!(m-2t-l\prime)!}\left(\frac{ik}{2z}\rho_y\right)^{m-2t-l\prime} \left(\frac{1}{2\sigma_y^2}+\frac{1}{\Lambda^2}\right)^{l\prime} \\
& \sqrt{\frac{\pi}{b_y}} 2^{-(n+l\prime)} i^{n+l\prime} \exp\left(\frac{c_y^2}{b_y}\right)\left(\frac{1}{b_y}\right)^{0.5(n+l\prime)} H_{n+l\prime}\left(\frac{ic_y}{\sqrt{b_y}}\right)
\end{aligned}
\tag{12}
$$

$$
\begin{aligned}
I_5 = {} & \sqrt{\frac{\pi}{a_x}}(M-m)!\left(\frac{1}{a_x}\right)^{M-m} \exp\left[\frac{1}{a_x}\left(\frac{ik}{2z}\rho_x\right)^2\right] \\
& \sum_{k=0}^{[\frac{M-m}{2}]} \frac{1}{k!(M-m-2k)!}\left(\frac{a_x}{4}\right)^k \sum_{l=0}^{[M-m-2k]} \frac{(M-m-2k)!}{l!(M-m-2k-l)!}\left(\frac{ik}{2z}\rho_x\right)^{M-m-2k-l} \\
& \left(\frac{1}{2\sigma_x^2}+\frac{1}{\Lambda^2}\right)^l \sqrt{\frac{\pi}{b_x}} 2^{-(M-n+l)} i^{M-n+l} \exp\left(\frac{c_x^2}{b_x}\right)\left(\frac{1}{b_x}\right)^{0.5(M-n+l)} H_{M-n+l}\left(\frac{ic_x}{\sqrt{b_x}}\right) \\
& \sqrt{\frac{\pi}{a_y}} m!\left(\frac{1}{a_y}\right)^m \exp\left[\frac{1}{a_y}\left(\frac{ik}{2z}\rho_y\right)^2\right] \\
& \sum_{t=0}^{[\frac{m}{2}]} \frac{1}{t!(m-2t)!}\left(\frac{a_y}{4}\right)^t \sum_{l\prime=0}^{[m-2t]} \frac{(m-2t)!}{l\prime!(m-2t-l\prime)!}\left(\frac{ik}{2z}\rho_y\right)^{m-2t-l\prime}\left(\frac{1}{2\sigma_y^2}+\frac{1}{\Lambda^2}\right)^{l\prime} \\
& \sqrt{\frac{\pi}{b_y}} 2^{-(n+2+l\prime)} i^{n+2+l\prime} \exp\left(\frac{c_y^2}{b_y}\right)\left(\frac{1}{b_y}\right)^{0.5(n+2+l\prime)} H_{n+l\prime}\left(\frac{ic_y}{\sqrt{b_y}}\right)
\end{aligned} \tag{13}
$$

$$
\begin{aligned}
I_6 = {} & \sqrt{\frac{\pi}{a_x}}(M-m+2)!\left(\frac{1}{a_x}\right)^{M-m+2} \exp\left[\frac{1}{a_x}\left(\frac{ik}{2z}\rho_x\right)^2\right] \\
& \sum_{k=0}^{[\frac{M-m+2}{2}]} \frac{1}{k!(M-m+2-2k)!}\left(\frac{a_x}{4}\right)^k \sum_{l=0}^{[M-m+2-2k]} \frac{(M-m+2-2k)!}{l!(M-m+2-2k-l)!}\left(\frac{ik}{2z}\rho_x\right)^{M-m+2-2k-l} \\
& \left(\frac{1}{2\sigma_x^2}+\frac{1}{\Lambda^2}\right)^l \sqrt{\frac{\pi}{b_x}} 2^{-(M-n+2+l)} i^{M-n+2+l} \exp\left(\frac{c_x^2}{b_x}\right)\left(\frac{1}{b_x}\right)^{0.5(M-n+2+l)} H_{M-n+2+l}\left(\frac{ic_x}{\sqrt{b_x}}\right) \\
& \sqrt{\frac{\pi}{a_y}} m!\left(\frac{1}{a_y}\right)^m \exp\left[\frac{1}{a_y}\left(\frac{ik}{2z}\rho_y\right)^2\right] \\
& \sum_{t=0}^{[\frac{m}{2}]} \frac{1}{t!(m-2t)!}\left(\frac{a_y}{4}\right)^t \sum_{l\prime=0}^{[m-2t]} \frac{(m-2t)!}{l\prime!(m-2t-l\prime)!}\left(\frac{ik}{2z}\rho_y\right)^{m-2t-l\prime}\left(\frac{1}{2\sigma_y^2}+\frac{1}{\Lambda^2}\right)^{l\prime} \\
& \sqrt{\frac{\pi}{b_y}} 2^{-(n+l\prime)} i^{n+l\prime} \exp\left(\frac{c_y^2}{b_y}\right)\left(\frac{1}{b_y}\right)^{0.5(n+l\prime)} H_{n+l\prime}\left(\frac{ic_y}{\sqrt{b_y}}\right)
\end{aligned} \tag{14}
$$

$$
\begin{aligned}
I_7 = {} & \sqrt{\frac{\pi}{a_x}}(M-m+2)!\left(\frac{1}{a_x}\right)^{M-m+2} \exp\left[\frac{1}{a_x}\left(\frac{ik}{2z}\rho_x\right)^2\right] \\
& \sum_{k=0}^{[\frac{M-m+2}{2}]} \frac{1}{k!(M-m+2-2k)!}\left(\frac{a_x}{4}\right)^k \sum_{l=0}^{[M-m+2-2k]} \frac{(M-m+2-2k)!}{l!(M-m+2-2k-l)!}\left(\frac{ik}{2z}\rho_x\right)^{M-m+2-2k-l} \\
& \left(\frac{1}{2\sigma_x^2}+\frac{1}{\Lambda^2}\right)^l \sqrt{\frac{\pi}{b_x}} 2^{-(M-n+l)} i^{M-n+l} \exp\left(\frac{c_x^2}{b_x}\right)\left(\frac{1}{b_x}\right)^{0.5(M-n+l)} H_{M-n+l}\left(\frac{ic_x}{\sqrt{b_x}}\right) \\
& \sqrt{\frac{\pi}{a_y}} m!\left(\frac{1}{a_y}\right)^m \exp\left[\frac{1}{a_y}\left(\frac{ik}{2z}\rho_y\right)^2\right] \\
& \sum_{t=0}^{[\frac{m}{2}]} \frac{1}{t!(m-2t)!}\left(\frac{a_y}{4}\right)^t \sum_{l\prime=0}^{[m-2t]} \frac{(m-2t)!}{l\prime!(m-2t-l\prime)!}\left(\frac{ik}{2z}\rho_y\right)^{m-2t-l\prime}\left(\frac{1}{2\sigma_y^2}+\frac{1}{\Lambda^2}\right)^{l\prime} \\
& \sqrt{\frac{\pi}{b_y}} 2^{-(n+2+l\prime)} i^{n+l\prime} \exp\left(\frac{c_y^2}{b_y}\right)\left(\frac{1}{b_y}\right)^{0.5(n+2+l\prime)} H_{n+l\prime}\left(\frac{ic_y}{\sqrt{b_y}}\right)
\end{aligned} \tag{15}
$$

$$\begin{aligned}
I_8 = &\sqrt{\frac{\pi}{a_x}}(M-m)!\left(\frac{1}{a_x}\right)^{M-m}\exp\left[\frac{1}{a_x}\left(\frac{ik}{2z}\rho_x\right)^2\right] \\
&\sum_{k=0}^{[\frac{M-m}{2}]}\frac{1}{k!(M-m-2k)!}\left(\frac{a_x}{4}\right)^k\sum_{l=0}^{[M-m-2k]}\frac{(M-m-2k)!}{l!(M-m-2k-l)!}\left(\frac{ik}{2z}\rho_x\right)^{M-m-2k-l} \\
&\left(\frac{1}{2\sigma_x^2}+\frac{1}{\Lambda^2}\right)^l\sqrt{\frac{\pi}{b_x}}2^{-(M-n+2+l)}i^{M-n+2+l}\exp\left(\frac{c_x^2}{b_x}\right)\left(\frac{1}{b_x}\right)^{0.5(M-n+2+l)}H_{M-n+l}\left(\frac{ic_x}{\sqrt{b_x}}\right) \\
&\sqrt{\frac{\pi}{a_y}}(m+2)!\left(\frac{1}{a_y}\right)^{m+2}\exp\left[\frac{1}{a_y}\left(\frac{ik}{2z}\rho_{1y}-\frac{\rho_{1y}-\rho_{2y}}{2\Lambda^2}\right)^2\right] \\
&\sum_{t=0}^{[\frac{m+2}{2}]}\frac{1}{t!(m+2-2t)!}\left(\frac{a_y}{4}\right)^t\sum_{l\prime=0}^{[m+2-2t]}\frac{(m+2-2t)!}{l\prime!(m+2-2t-l\prime)!}\left(\frac{ik}{2z}\rho_y\right)^{m+2-2t-l\prime}\left(\frac{1}{2\sigma_y^2}+\frac{1}{\Lambda^2}\right)^{l\prime} \\
&\sqrt{\frac{\pi}{b_y}}2^{-(n+l\prime)}i^{n+l\prime}\exp\left(\frac{c_y^2}{b_y}\right)\left(\frac{1}{b_y}\right)^{0.5(n+l\prime)}H_{n+l\prime}\left(\frac{ic_y}{\sqrt{b_y}}\right)
\end{aligned} \quad (16)$$

$$\begin{aligned}
I_9 = &\sqrt{\frac{\pi}{a_x}}(M-m)!\left(\frac{1}{a_x}\right)^{M-m}\exp\left[\frac{1}{a_x}\left(\frac{ik}{2z}\rho_x\right)^2\right] \\
&\sum_{k=0}^{[\frac{M-m}{2}]}\frac{1}{k!(M-m-2k)!}\left(\frac{a_x}{4}\right)^k\sum_{l=0}^{[M-m-2k]}\frac{(M-m-2k)!}{l!(M-m-2k-l)!}\left(\frac{ik}{2z}\rho_x\right)^{M-m-2k-l} \\
&\left(\frac{1}{2\sigma_x^2}+\frac{1}{\Lambda^2}\right)^l\sqrt{\frac{\pi}{b_x}}2^{-(M-n+l)}i^{M-n+l}\exp\left(\frac{c_x^2}{b_x}\right)\left(\frac{1}{b_x}\right)^{0.5(M-n+l)}H_{M-n+l}\left(\frac{ic_x}{\sqrt{b_x}}\right) \\
&\sqrt{\frac{\pi}{a_y}}(m+2)!\left(\frac{1}{a_y}\right)^{m+2}\exp\left[\frac{1}{a_y}\left(\frac{ik}{2z}\rho_y\right)^2\right] \\
&\sum_{t=0}^{[\frac{m+2}{2}]}\frac{1}{t!(m+2-2t)!}\left(\frac{a_y}{4}\right)^t\sum_{l\prime=0}^{[m+2-2t]}\frac{(m+2-2t)!}{l\prime!(m+2-2t-l\prime)!}\left(\frac{ik}{2z}\rho_y\right)^{m+2-2t-l\prime}\left(\frac{1}{2\sigma_y^2}+\frac{1}{\Lambda^2}\right)^{l\prime} \\
&\sqrt{\frac{\pi}{b_y}}2^{-(n+2+l\prime)}i^{n+2+l\prime}\exp\left(\frac{c_y^2}{b_y}\right)\left(\frac{1}{b_y}\right)^{0.5(n+2+l\prime)}H_{n+2+l\prime}\left(\frac{ic_y}{\sqrt{b_y}}\right)
\end{aligned} \quad (17)$$

with

$$a_\beta = \frac{1}{w_\beta^2}+\frac{1}{2\sigma_\beta^2}+\frac{1}{\Lambda^2}+\frac{ik}{2z}(\beta = x, y) \quad (18)$$

$$b_\beta = \frac{1}{w_\beta^2}+\frac{1}{2\sigma_\beta^2}+\frac{1}{\Delta^2}-\frac{ik}{2z}-\frac{1}{b_\beta}\left(\frac{1}{2\sigma_\beta^2}+\frac{1}{\Lambda^2}\right) \quad (19)$$

$$c_\beta = \frac{ik}{2z}\rho_\beta-\frac{ik}{2a_\beta z}\left(\frac{1}{2\sigma^2}+\frac{1}{\Lambda^2}\right)\rho_\beta \quad (20)$$

By using Equations (8)–(20), the evolution of intensity of PCAAVLB in turbulent biological tissue can be obtained.

3. Results and Analyses

In this section, the intensity of a PCAAVLB in turbulent biological tissue is analyzed based on the numerical simulations. In the numerical simulation, the following parameter are chosen, such as $\lambda = 0.8$ μm, $w_{0x} = w_{0y} = 20$ μm, $\sigma_x = \sigma_y = \sigma = 5$ μm, $M = 1$, $S = 10^{-4}$, $l_c = 10$ μm, $l_0 = 1$ μm, and $D = 4$.

First, the intensity of a PCAAVLB with $a = 2$, $c_x = c_y = 8$, and $\sigma_x = \sigma_y = 5$ μm in free space (Λ = inf) is illustrated in Figure 1. The intensity of this PCAAVLB in free space shows a ring profile at $z = 20$ μm (Figure 1a). As z increases, the hollow width of ring shape of such a PCAAVLB will gradually lessen (Figure 1b) and disappear (Figure 1c).

At last, the intensity of this PCAAVLB will become a spot pattern with a Gaussian-like profile (Figure 1d). Then the adjustable parameters change. Figure 2 shows the intensity of a PCAAVLB with $a = 1$, $c_x = 2$, and $c_y = 8$ in free space. When $c_x \neq c_y$, the intensity of this PCAAVLB can show the two-spot pattern at z = 20 μm (Figure 2a). The PCAAVLB with $c_x = c_y$ has the ring profile (Figure 1a). As z increases, the two-spot pattern of this PCAAVLB will overlap (Figure 2b,c). Lastly, the intensity of this PCAAVLB can become an elliptical profile (Figure 2d) when $c_x > c_y$; the intensity of PCAAVLB with $a = 1$, $c_x = 8$, and $c_y = 2$ in free space is shown in Figure 3. The intensity of such a PCAAVLB will have two spots along x-axis (Figure 3a), while the PCAAVLB with $c_x < c_y$ shows two spots along the y-axis (Figure 2a). And this PCAAVLB will evolve into an elliptical beam along the x-axis (Figure 3b). Therefore, the intensity of such PCAAVLB can be controlled by adjustable parameters and the circular or elliptical intensity shape can be obtained by setting the different parameters.

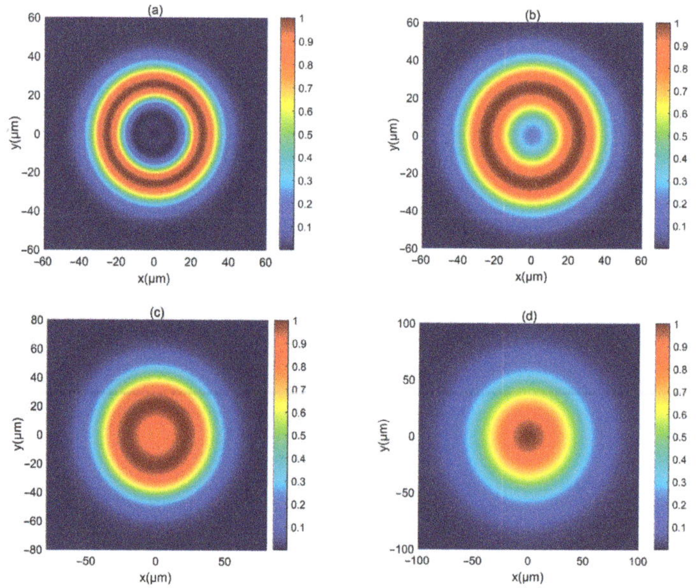

Figure 1. The normalized intensity of a PCAAVLB with $a = 2$, $c_x = c_y = 8$ in free space: (**a**) z = 20 μm, (**b**) z = 300 μm, (**c**) z = 500 μm, and (**d**) z = 1000 μm.

Next, the intensity of such PCAAVLB in turbulent biological tissue is studied. Figure 4 illustrates the intensity of a PCAAVLB with $a = 2$ and $c_x = c_y = 8$ in turbulent biological tissue. Intensity shape of this PCAAVLB in turbulent biological tissue at z = 20 μm retains its ring shape (Figure 4a), and the hollow center will gradually disappear at z = 150 μm (Figure 4b), while the PCAAVLB with δ = 5 μm in free space at z = 300 μm just has the ring shape (Figure 1a). Therefore, the biological tissue will accelerate the speed of the PCAAVLB ring shape loss. As z increases, this PCAAVLB in turbulent biological tissue will show a Gaussian-like shape (Figure 4c), and the spot of intensity will spread farther (Figure 4d). When $c_x \neq c_y$, the intensity of PCAAVLB with $c_x < c_y$ in turbulent biological tissue becomes an elliptical profile with the long axis along the y-axis at z = 300 μm (Figure 5a), while the same PCAAVLB in free space will evolve into a two-spot pattern (Figure 2c). In addition, the intensity of PCAAVLB with $c_x > c_y$ in turbulent biological tissue will evolve into an elliptical Gaussian profile with the long axis along the x-axis at z = 300 μm (Figure 5b).

From Figures 4 and 5, one concludes that the PCAAVLB in turbulent biological tissue will evolve into one spot pattern faster as z increases.

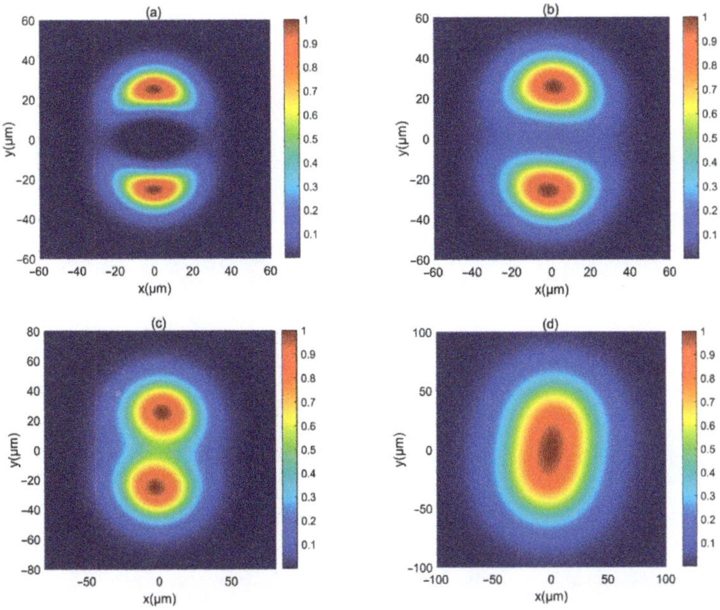

Figure 2. The normalized intensity of a PCAAVLB with $a = 1$, $c_x = 2$, and $c_y = 8$ in free space: (**a**) z = 20 μm, (**b**) z = 300 μm, (**c**) z = 500 μm, and (**d**) z = 1000 μm.

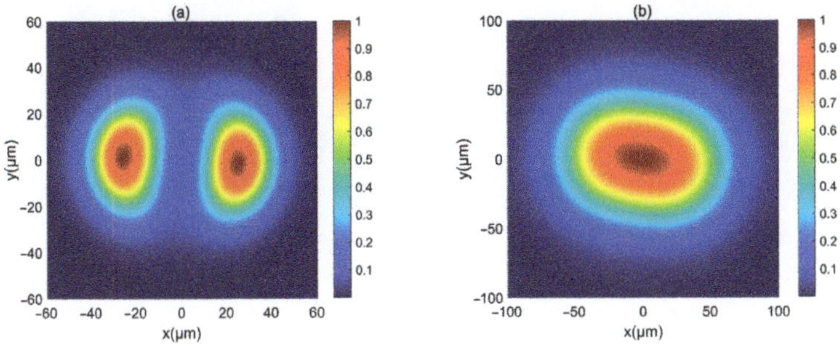

Figure 3. The normalized intensity of a PCAAVLB with $a = 1$ and $c_x = 8$, and $c_y = 2$ in free space: (**a**) z = 300 μm, and (**b**) z = 1000 μm.

To see the beam parameters on the intensity of a PCAAVLB in turbulent biological tissue, Figure 6 shows the intensity $I(r,z)/I_{\max}(r,0)$ of a PCAAVLB with $a = -4$, $c_x = c_y = 8$, and $M = 2$ in biological tissue for the different σ. The intensity of PCAAVLBs with the different σ in turbulent biological tissue has the same profile at $z = 20$ μm (Figure 6a). As z increases, such PCAAVLB with smaller σ in turbulent biological tissue will first lose the ring profile (Figure 6a) and gradually become a Gaussian-like profile (Figure 6c). As z increases further, such PCAAVLBs with the different σ can all have Gaussian-like patterns, such PCAAVLBs with smaller σ will have a larger width, and the intensity of PCAAVLB with smaller σ will decrease faster (Figure 6d). Thus, the speed of evolution can be controlled by σ, and the PCAAVLBs with a smaller σ will evolve into a Gaussian shape faster and have a smaller intensity.

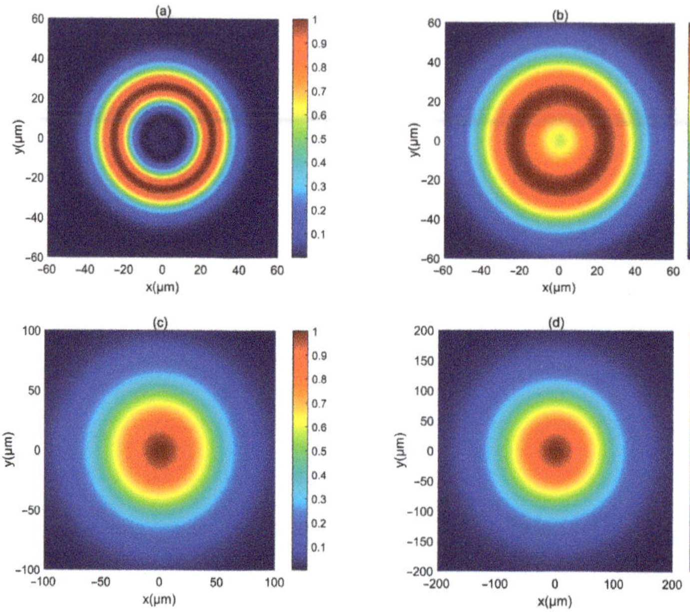

Figure 4. The normalized intensity of a PCAAVLB with $a = 2$ and $c_x = c_y = 8$ in biological tissue for the different δ: (**a**) z = 20 μm, (**b**) z = 150 μm, (**c**) z = 300 μm, and (**d**) z = 500 μm.

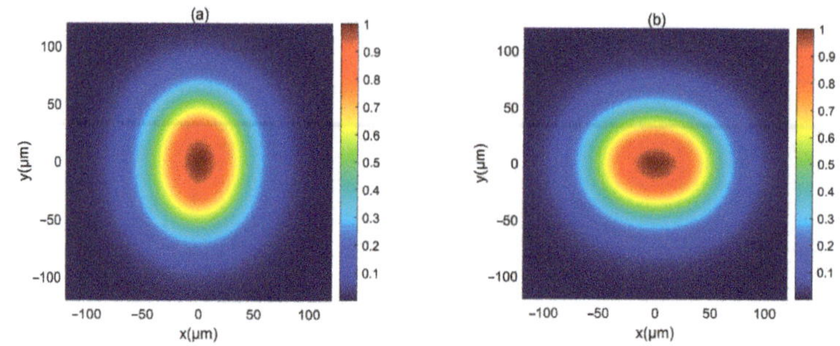

Figure 5. Cross sections of a PCAAVLB with $a = -2$ and $c_x = c_y = 8$ in biological tissue at z = 300 μm: (**a**) $a = 1$, $c_x = 2$, and $c_y = 8$; (**b**) $a = 1$, $c_x = 8$, and $c_y = 2$.

The intensity $\mathrm{I}(r,z)/I_{\max}(r,0)$ of a PCAAVLB with $a = -4$, $c_x = c_y = 8$, and $\sigma = 2$ μm in turbulent biological tissue for the different M is illustrated in Figure 7. The PCAAVLB with a larger M has a bigger width at $z = 100$ μm (Figure 7a). As z increases, the PCAAVLB with a smaller M in turbulent biological tissue will first lose its ring profile, while the intensity of the PCAAVLB with a larger M will have a lower intensity (Figure 7b). As z increases further, the PCAAVLB with a different M can gradually become a Gaussian profile (Figure 7c), and such a PCAAVLB with a smaller M in turbulent biological tissue will have a larger intensity although all PCAAVLBs have the Gaussian profile (Figure 7d).

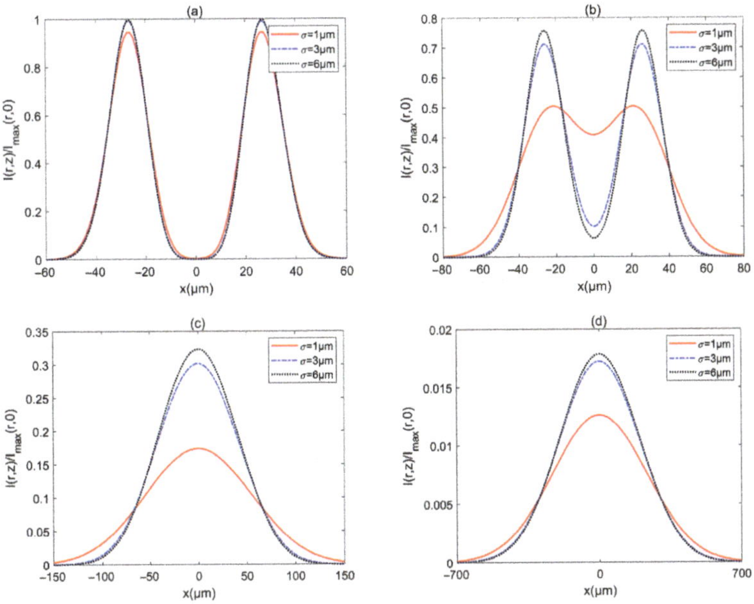

Figure 6. Cross sections of intensity of a PCAAVLB with $a = -4$ and $c_x = c_y = 8$ in biological tissue for the different δ: (**a**) z = 20 μm, (**b**) z = 100 μm, (**c**) z = 300 μm, and (**d**) z = 1000 μm.

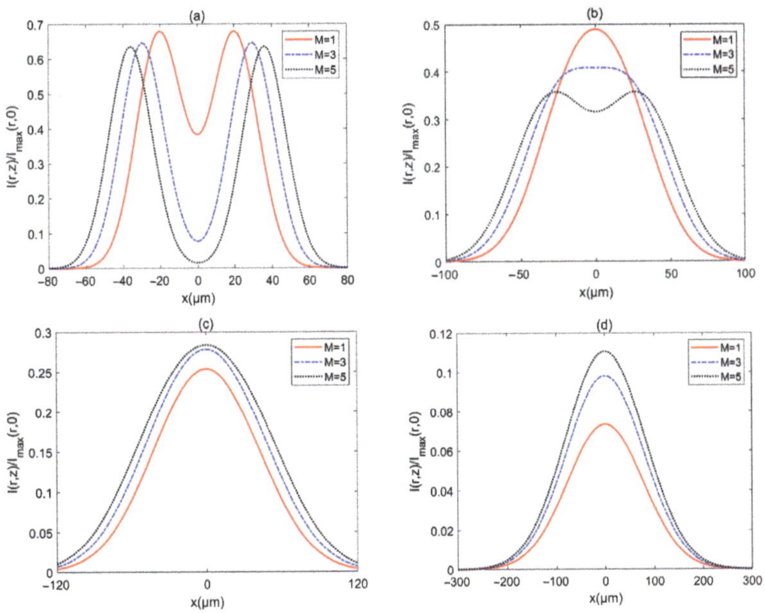

Figure 7. Cross sections of intensity of a PCAAVLB with $a = -4$ and $c_x = c_y = 8$ in biological tissue for the different M: (**a**) z = 100 μm, (**b**) z = 200 μm, (**c**) z = 300 μm, and (**d**) z = 500 μm.

The influences of parameters of turbulent biological tissue on the intensity of PCAAVLB are shown in Figures 8 and 9. The intensity $I(r,z)/I_{\max}(r,0)$ of a PCAAVLB with $a = -4$, $c_x = c_y = 8$, and $\sigma = 2$ μm in turbulent biological tissue for the different D is illustrated in Figure 8. The intensity profile of such a PCAAVLB in turbulent biological tissue with different D remains the ring profile at $z = 100$ μm (Figure 8a). As z increases, such PCAAVLB in a turbulent biological tissue with a larger D will first lose its ring shape (Figure 8b), evolve into a Gaussian profile, and the beam in a turbulent biological tissue with a larger D will have a larger intensity. (Figure 8c). The PCAAVLB in a turbulent biological tissue with a larger D will have a larger spot at $z = 500$ μm and remain the larger intensity (Figure 8d).

Figure 9 shows the intensity $I(r,z)/I_{max}(r,0)$ of a PCAAVLB with $a = -4$, $c_x = c_y = 8$, and $\sigma = 2$ μm in turbulent biological tissue for different l_0. The intensity profile of such a PCAAVLB in turbulent biological tissue with different l_0 remains almost similar at z = 100 x3BCm (Figure 9a). As z increases, the PCAAVLB in turbulent biological tissue for the different l_0 will have similar evolution properties (Figure 9b,c). At last, the PCAAVLB in turbulent biological tissue with different l_0 will have a similar spot and the beam with larger l_0 will have a larger intensity. Therefore, one can conclude that the PCAAVLB in turbulent biological tissue with a larger D or a smaller l_0 will evolve from a ring profile into a Gaussian-like profile faster, and such a PCAAVLB in turbulent biological tissue with a larger D or a smaller l_0 will have a larger intensity.

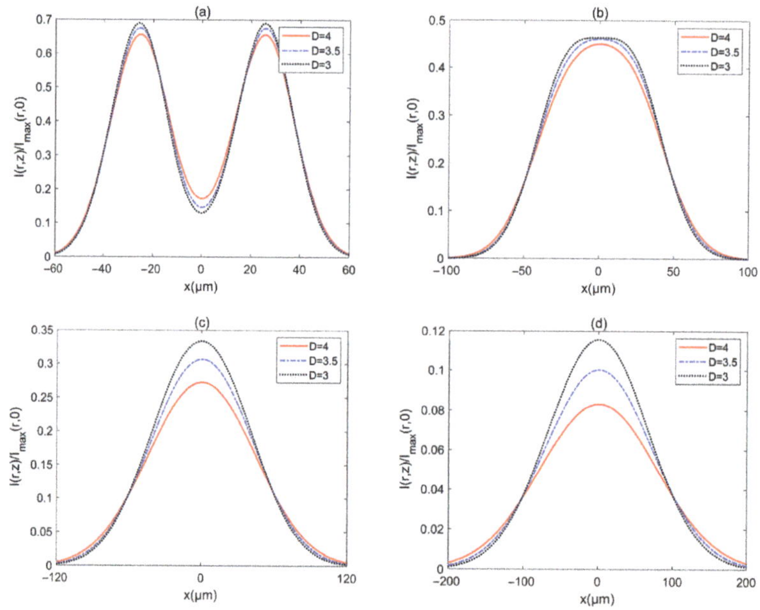

Figure 8. Cross sections of intensity of a PCAAVLB with $a = -4$ and $c_x = c_y = 8$ in biological tissue for the different D: (**a**) z = 100 μm, (**b**) z = 200 μm, (**c**) z = 300 μm, and (**d**) z = 500 μm.

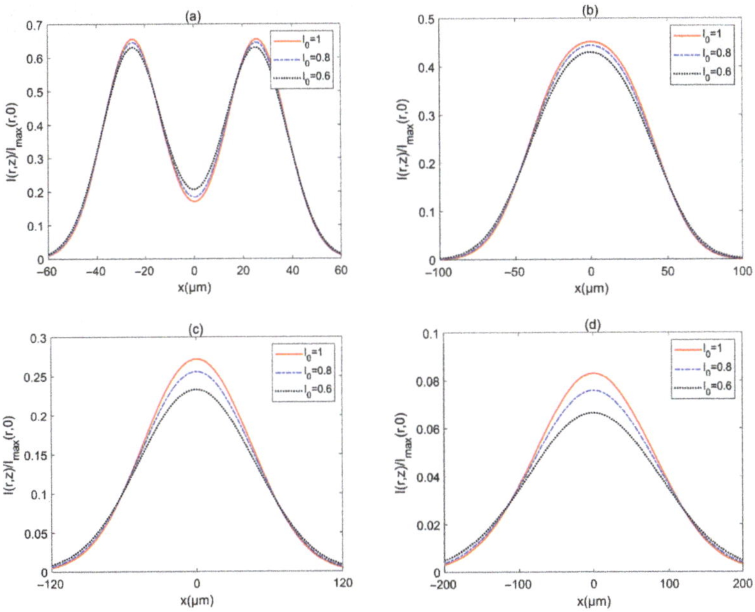

Figure 9. Cross sections of intensity of a PCAAVLB with $a = -4$ and $c_x = c_y = 8$ in biological tissue for the different l_0: (**a**) z = 100 μm, (**b**) z = 200 μm, (**c**) z = 300 μm, and (**d**) z = 500 μm.

4. Conclusions

Vortices are ubiquitous in nature, such as in fluids, smoke rings, and tornados, and vortices have also been observed in light. In this work, a new beam carried vortex, which is named PCAAVLB, is introduced, and the intensity of such PCAAVLB in turbulent biological tissue is derived. The intensity shape of such PCAAVLB in free space can have the ring shape or a two-spot pattern, and the intensity profile is controlled by adjustable initial parameters. The intensity of such a PCAAVLB can gradually evolve into a Gaussian-like profile. And the intensity of such a PCAAVLB with a smaller M or σ in turbulent biological tissue will lose its ring shape faster, and the intensity profile of such a PCAAVLB with smaller σ in turbulent biological tissue will have a larger spot, while the PCAAVLB with different M in turbulent biological tissue has almost the same intensity profile at a larger distance. In addition, the PCAAVLB in turbulent biological tissue with larger D or smaller l_0 will evolve from a ring profile into a Gaussian-like profile faster and have a larger intensity. The obtained results may have potential applications in laser sensing under biological tissue environments and laser imaging in biology.

Author Contributions: Conceptualization, Y.S. and R.D.; software, W.W. and X.D.; data curation, L.S. and Z.Q.; writing—original draft preparation, Y.Z.; writing—review and editing, H.J.; project administration, Y.S. All authors have read and agreed to the published version of the manuscript.

Funding: This work was supported by the National Key R&D Program Project 'Intersatellite Laser Communication Measurement Integrated System Technology' (2022YFB3902505), the National Natural Science Foundation of China (under Grant 62201307), the 111 Project of China (D21009), and the Major Key Project of PCL (PCL2024A01).

Institutional Review Board Statement: Not applicable.

Informed Consent Statement: Not applicable.

Data Availability Statement: The data presented in this work are available upon request from the corresponding author.

Conflicts of Interest: Author Wu Wang was employed by the North Navigation Control Technology Co., Ltd. The remaining authors declare that the research was conducted in the absence of any commercial or financial relationships that could be construed as a potential conflict of interest.

References

1. Jacques, S.L. Optical properties of biological tissues: A review. *Phys. Med. Biol.* **2013**, *58*, R37–R61. [CrossRef] [PubMed]
2. Gökçe, M.C.; Ata, Y.; Baykal, Y. Tissue turbulence and its effects on optical waves: A review. *Opt. Commun.* **2023**, *546*, 129816. [CrossRef]
3. Saad, F.; Belafhal, A. A theoretical investigation on the propagation properties of Hollow Gaussian beams passing through turbulent biological tissues. *Optik* **2017**, *141*, 72–82. [CrossRef]
4. Wu, Y.Q.; Zhang, Y.X.; Wang, Q.; Hu, Z.D. Average intensity and spreading of partially coherent model beams propagating in a turbulent biological tissue. *J. Quant. Spectrosc. Radiat. Transf.* **2016**, *184*, 308–315. [CrossRef]
5. Liu, D.; Zhong, H.; Wang, Y. Intensity properties of anomalous hollow vortex beam propagating in biological tissues. *Optik* **2018**, *170*, 61–69. [CrossRef]
6. Zhang, H.; Cui, Z.; Han, Y.; Guo, J.; Chang, C. Average Intensity and Beam Quality of Hermite-Gaussian Correlated Schell-Model Beams Propagating in Turbulent Biological Tissue. *Front. Phys.* **2021**, *9*, 650537. [CrossRef]
7. Liu, D.J.; Zhong, H.Y.; Yin, H.M.; Dong, A.Y.; Wang, G.Q.; Wang, Y.C. Spreading and coherence properties of a rectangular multi-Gaussian Schell-model beam propagating in biological tissues. *Indian J. Phys.* **2021**, *95*, 571–577. [CrossRef]
8. Duan, M.; Tian, Y.; Zhang, Y.; Li, J. Influence of biological tissue and spatial correlation on spectral changes of Gaussian-Schell model vortex beam. *Opt. Laser. Eng.* **2020**, *134*, 106224. [CrossRef]
9. Duan, M.L.; Wu, Y.G.; Zhang, Y.M.; Li, J.H. Coherence properties of a random electromagnetic vortex beam propagating in biological tissues. *J. Mod. Optic.* **2019**, *66*, 59–66. [CrossRef]
10. Liang, Q.; Hu, B.; Zhang, Y.; Zhu, Y.; Deng, S.; Yu, L. Coupling efficiency of a partially coherent collimating laser from turbulent biological tissue to fiber. *Results Phys.* **2019**, *13*, 102162. [CrossRef]
11. Baykal, Y. Adaptive optics correction of beam spread in biological tissues. *J. Quant. Spectrosc. Radiat. Transf.* **2022**, *283*, 108145. [CrossRef]
12. Dorrah, A.H.; Palmieri, A.; Li, L.; Capasso, F. Rotatum of light. *Sci. Adv.* **2025**, *11*, eadr9092. [CrossRef] [PubMed]
13. Gowree, E.R.; Jagadeesh, C.; Talboys, E.; Lagemann, C.; Brücker, C. Vortices enable the complex aerobatics of peregrine falcons. *Commun. Biol.* **2018**, *1*, 27. [CrossRef]
14. Bomphrey, R.J.; Nakata, T.; Phillips, N.; Walker, S.M. Smart wing rotation and trailing-edge vortices enable high frequency mosquito flight. *Nature* **2017**, *544*, 92–95. [CrossRef] [PubMed]
15. Guo, M.; Le, W.; Wang, C.; Rui, G.; Zhu, Z.; He, J.; Gu, B. Generation, Topological Charge, and Orbital Angular Momentum of Off-Axis Double Vortex Beams. *Photonics* **2023**, *10*, 368. [CrossRef]
16. Chen, L.; Liu, D.; Gao, H.; Dong, A.; Wang, Y. Research on characteristics of partially coherent radially polarized off-axis double vortex beam in oceanic turbulence. *Opt. Laser Technol.* **2025**, *190*, 113264. [CrossRef]
17. Cai, Y.J. Model for an anomalous hollow beam and its paraxial propagation. *Opt. Lett.* **2007**, *32*, 3179–3181. [CrossRef]
18. Dai, Z.P.; Wang, Y.J.; Wang, Y.C.; Liu, X.J.; Yang, Z.J.; Pang, Z.G. Propagation of adjustable anomalous hollow Gaussian beams in the cascade of left-handed and right-handed media. *Optik* **2022**, *270*, 170024. [CrossRef]
19. Dai, Z.P.; Wang, Y.J.; Wang, Y.C.; Yang, Z.J.; Pang, Z.G. Propagation properties of controllable anomalous hollow beams in strongly nonlocal nonlinear media. *Results Phys.* **2022**, *42*, 105951. [CrossRef]
20. Dai, Z.P.; Yang, Z.J.; Zhang, S.M.; Pang, Z.G.; You, K.M. Transversal reverse transformation of anomalous hollow beams in strongly isotropic nonlocal media. *Chin. Phys. B* **2014**, *23*, 074208. [CrossRef]
21. Wang, K.L.; Zhao, C.H. Propagation properties of a radial phased-locked partially coherent anomalous hollow beam array in turbulent atmosphere. *Opt. Laser Technol.* **2014**, *57*, 44–51. [CrossRef]
22. Tian, H.H.; Xu, Y.G.; Yang, T.; Ma, Z.R.; Wang, S.J.; Dan, Y.Q. Propagation characteristics of partially coherent anomalous elliptical hollow Gaussian beam propagating through atmospheric turbulence along a slant path. *J. Mod. Optic.* **2017**, *64*, 422–429. [CrossRef]
23. Liu, D.; Wang, Y.; Zhong, H.; Wang, G.; Yin, H.; Dong, A. Properties of multi-Gaussian correlated partially coherent anomalous hollow beam propagating in underwater oceanic turbulence. *Opt. Laser Technol.* **2019**, *119*, 105604. [CrossRef]
24. Guo, X.; Song, Y.; Dong, K.; Zhang, L. Adjustable anomalous hollow vortex beam and its properties. *Microw. Opt. Techn. Let.* **2024**, *66*, e34099. [CrossRef]

25. Wang, Y.J.; Yang, Z.J.; Shen, S.; Cai, Z.H.; Li, X.L.; Zhang, S.M. Controllable anomalous hollow beam and its propagation properties. *Optik* **2021**, *241*, 167054. [CrossRef]
26. Wolf, E. Unified theory of coherence and polarization of random electromagnetic beams. *Phys. Lett. A* **2003**, *312*, 263–267. [CrossRef]
27. Jeffrey, A. *Handbook of Mathematical Formulas and Integrals*, 4th ed.; Academic Press Inc.: Cambridge, MA, USA, 2008.

MDPI AG
Grosspeteranlage 5
4052 Basel
Switzerland
Tel.: +41 61 683 77 34

Biomimetics Editorial Office
E-mail: biomimetics@mdpi.com
www.mdpi.com/journal/biomimetics

www.ingramcontent.com/pod-product-compliance
Lightning Source LLC
Chambersburg PA
CBHW062045220926
51850CB00010B/2664